ALMOST COMPLEX HOMOGENEOUS SPACES AND THEIR SUBMANIFOLDS

ALMOST COMPLEX HOMOGENEOUS SPACES AND THEIR SUBMANIFOLDS

KICHOON YANG

Arkansas State University

World Scientific
Singapore • New Jersey • Hong Kong

Published by

World Scientific Publishing Co. Pte. Ltd.
P.O. Box 128, Farrer Road, Singapore 9128

U. S. A. office: World Scientific Publishing Co., Inc.
687 Hartwell Street, Teaneck NJ 07666, USA

Library of Congress Cataloging-in-Publication data is available.

ISBN 9971-50-377-8

Printed in Singapore by Lenco Printing Co. Pte. Ltd.

PREFACE

These notes contain an exposition of the theory of almost complex homogeneous spaces and their complex submanifolds. Certain closely related class of spaces, what we call partial G-flag manifolds, are also studied. Our emphasis is on the compact case so that there is a pervasive influence of the structure theory of compact Lie groups. In this we are strongly guided by the works of Wang [Wan] and Borel-Hirzebruch [B-H].

An almost complex homogeneous space is a homogeneous space equipped with an invariant almost complex structure. In the compact case fundamental examples of such spaces are provided by homogeneous spaces of the form G/H, where G is a connected compact semisimple Lie group and H is a closed subgroup of maximal rank. (Not all such homogeneous spaces are almost complex, however. We call such a space a partial G-flag manifold.) For these spaces the root decomposition of G relative to a maximal torus T (contained in H) plays a crucial role and the understanding of the flag manifold G/T becomes important.

Recently much attention has been paid to the investigation of minimal surfaces (and harmonic maps) in various Riemannian manifolds. Often times in these studies the ambient space has been an almost complex homogeneous space or at least a partial G-flag manifold and when this happened there appeared a strong relationship between (pseudo-) complex curves and minimal surfaces. Just to mention a few examples we have $CP^n = U(n+1)/U(1) \times U(n)$ ([C-W], [E-W]), $HP^n = Sp(n+1)/Sp(1) \times Sp(n)$ ([G1]), $Q_n = SO(n+2;\mathbf{R})/SO(2) \times SO(n)$ ([J-R-Y]), $S^4 = HP^1$ ([Ch1], [B2]), $S^6 = G_2/SU(3)$ ([B1]). No doubt that a systematic investigation into (pseudo-) complex submanifolds of these spaces is worth pursuing.

The method we adopt, justifiably most suited to our situation, is that of moving frames (repère mobile) used so successfully by Cartan and more recently by Chern. Accordingly we reformulate the basic notions regarding invariant almost complex structures on G/H in terms of the Maurer-Cartan form of G which plays an essential role in studying submanifolds of homogeneous spaces via moving frames. The basic reference on the technique of moving frames is Cartan [Cal]. One may also consult [Gr] or [J] for a more recent treatment.

Moving frame theoretic analysis of a given class of submanifolds in a homogeneous space is often complicated by algebraic difficulties associated with computing certain orbit structures of Lie group actions. This problem is already formidable at local level in many instances. (Holomorphic curves in CP^n case is a notable exception to this. Here one is able to concentrate on the global study.) However, in studying complex submanifolds of a flag manifold the "local-algebraic" difficulty disappears. Purely algebraic consideration into the root system of a compact Lie group G relative to a maximal torus T enables us to understand local orbit structures of complex maps rather completely. Thus we are free to pursue the global investigation of these objects in an attempt to mimic the well-known situation of holomorphic curves in CP^n (e.g., Plücker formulae, equidistribution theorems, etc).

Understanding of the flag manifold (G/T) case frequently throws light upon other more general cases. For example, if G/H is an almost complex homogeneous space with H, a closed subgroup of maximal rank (G, compact) then H contains a maximal torus of G. Therefore, by composing complex maps into G/T with the projection $G/T \rightarrow G/H$ we recover complex maps into G/H. Of particular importance among complex submanifolds of G/T in this regard are "horizontal" complex submanifolds. An important theorem regarding horizontal complex curves in G/T and minimal surfaces in G/H, under a setting parallel to ours, was proved by Bryant [B3].

Though we have not proved this in its generality, a sort of converse to the above statement about the projections of complex submanifolds of G/T to G/H is exhibited in Ch. IV §§18–19. This is achieved by interpreting the "Frenet frames" along complex submanifolds of G/H as horizontal complex liftings into G/T. To be sure there is more work to be done along this area.

Each chapter comes with its own introduction explaining the organization.

Arkansas State University
1987

Kichoon Yang

TABLE OF CONTENTS

Preface   v

Chapter I.   Structures of Compact Lie Groups   1
  1. Definitions and examples   1
  2. Lie algebras: basic results   4
  3. Orthogonal and unitary representations   9
  4. Maximal tori and Stiefel diagrams   12
  5. Weyl groups   18
  6. Dynkin diagrams and the classification   22

Chapter II.   Almost Complex Homogeneous Spaces   27
  7. Homogeneous spaces   27
  8. Partial G-flag manifolds and invariant
     metrics   30
  9. Invariant complex structures   32
  10. The Maurer-Cartan form   36
  11. Integrability condition   40
  12. The horizontal distribution   43

Chapter III.   Complex Submanifolds   45
  13. Pseudocomplex maps   45
  14. Moving frames   47
  15. $U(n)$-flag manifolds   50
  16. $SO(2n; \mathbf{R})$-flag manifolds   53

Chapter IV.   Examples: Holomorphic Curves   57
  17. Holomorphic curves in $F_{1,2,3}(\mathbf{C}^3)$ 58
  18. Horizontal curves in $U(n)/T$ 62

19. Holomorphic curves in CP^m — 67

20. Horizontal curves in $Sp(n)/T$ — 72

Chapter V.   Algebraic Curves — 79

21. Singular metrics and the
    Gauss-Bonnet-Chern theorem — 80

22. Projective curves and their
    associated curves — 82

23. Horizontal curves and the Frenet frames — 84

24. Plücker formulae for projective curves — 85

25. The symplectic Plücker formulae — 87

Appendix: Exterior Differential Systems — 91

1. Exterior algebra — 91

2. Completely integrable systems and
   the Cauchy characteristics — 93

3. Cartan-Kähler theory — 96

4. Prolongation — 99

References — 105

Index — 109

ALMOST COMPLEX
HOMOGENEOUS SPACES
AND THEIR SUBMANIFOLDS

Chapter I

STRUCTURES OF COMPACT LIE GROUPS

This chapter deals with the standard structure theory of compact Lie groups and is purely expository. Understanding of these materials on compact Lie groups is so crucial in further development of our theory that the author felt compelled to include them here. In addition it serves a useful purpose of setting up the notation for later chapters.

Our two main references for this chapter were [A] and [B-D]. We have also used the articles [A-B-S], [B1] and the books [He], [Hu] and [War].

§1. Definitions and Examples

In this section we give examples of (mostly compact) Lie groups as they naturally occur.

We begin with a theorem of Frobenius from algebra.

Theorem. Let K be a finite dimensional division algebra over $\mathbf{R}$ so that $\alpha x = x\alpha$ for every $\alpha \in \mathbf{R}$, $x \in K$. Then there are exactly three possibilities for K, i) $K = \mathbf{R}, \mathbf{C}$ (both commutative, hence fields), or ii) $k = \mathbf{H}$, the quaternions (noncommutative).

If we drop the associativity requirement of the multiplication in the ring structure of K then there is one other possibility, namely $\mathbf{O}$, the Cayley-Graves algebra, also called the octanians.

The division algebras admit another characterization as so called inner product algebras.

Definition. An inner product algebra K over $\mathbf{R}$ is a real vector space with a positive definite inner product $\langle\,,\,\rangle\colon K \times K \to \mathbf{R}$ and a

1

multiplication with 1 such that for every $x, y \in K$

$$\langle xy, xy \rangle = \langle x, x \rangle \langle y, y \rangle .$$

Theorem. There are exactly four inner product algebras over $\mathbf{R}$, and they are $\mathbf{R}, \mathbf{C}, \mathbf{H}, \mathbf{O}$.

For a proof see [B1] pp. 188–190 and the references cited there. Brief descriptions of $\mathbf{H}$ and $\mathbf{O}$ follow.

Construction of the Quaternions. $\mathbf{H}$ is defined as a four dimensional vector space over $\mathbf{R}$ with the basis $1, i, j, k$ and with the ring multiplication determined by

$$i^2 = j^2 = k^2 = -1, \ ij = k = -ji, \ jk = i = -kj, ki = j = -ik .$$

One may also think of $\mathbf{H}$ as a two dimensional left vector space over $\mathbf{C}$ with the basis $1, j$. So if $x = z_1 + z_2 j, y = \tilde{z}_1 + \tilde{z}_2 j$ are elements of $\mathbf{H}$ then we must have

$$xy = \left(z_1 \tilde{z}_1 + z_2 \bar{\tilde{z}}_2 \right) + \left(z_1 \tilde{z}_2 - z_2 \bar{\tilde{z}}_1 \right) j .$$

$\mathbf{H}$ also becomes a right $\mathbf{C}$-vector space by the rule $zj = j\bar{z}, z \in \mathbf{C}$. There is a natural conjugation in $\mathbf{H}$, denoted by *,

$$\left(z_1 + z_2 j \right)^* = \bar{z}_1 - z_2 j .$$

We define an inner product in $\mathbf{H}$ by

$$\langle x, y \rangle_{\mathbf{H}} = \operatorname{Re} xy^* = \frac{1}{2} \left(xy^* + yx^* \right) .$$

Construction of the Octanians. $\mathbf{O}$ is a two dimensional left vector space over $\mathbf{H}$ with the basis $1, \varepsilon$ and with the multiplication given by

$$\left(x + y\varepsilon \right) \left(\tilde{x} + \tilde{y}\varepsilon \right) = \left(x\tilde{x} - y\tilde{y}^* \right) + \left(x\tilde{y} + y\tilde{x}^* \right) \varepsilon ,$$

where $x, y, \tilde{x}, \tilde{y} \in \mathbf{H}$. From this its $\mathbf{R}$-algebra structure is determined. The inner product in $\mathbf{O}$ is given by the polarization of

$$\langle x + y\varepsilon, x + y\varepsilon \rangle_{\mathbf{O}} = xx^* + yy^* .$$

From now on we will confuse * with – whenever there is no real danger of confusion.

A Lie group is a real analytic manifold with a compatible group structure, i.e., $(x, y) \mapsto xy^{-1}$ is real analytic. A topologically closed subgroup of G inherits a Lie group structure.

Examples of Lie Groups.

i) $K = \mathbf{R}, \mathbf{C}, \mathbf{H}, \mathbf{O}$ are Lie groups under addition.

ii) $K^* = \mathbf{R}^*, \mathbf{C}^*, \mathbf{H}^*$ are Lie groups under multiplication, where K^* denotes nonzero elements of K.

iii) Let $K = \mathbf{R}, \mathbf{C}, \mathbf{H}$. Then we have the full matrix groups, denoted by $M(n; K)$, consisting of all $n \times n$ K-valued matrices. We also have the general linear groups, denoted by $GL(n; K)$, consisting of nonsingular matrices.

iv) There are the closed subgroups $O(n; \mathbf{R}) < GL(n; \mathbf{R}), U(n) < GL(n; \mathbf{C}), Sp(n) < GL(n; \mathbf{H})$ all defined by the condition ${}^t\bar{X}X = I$. In fact, let $V = K^n$ $(K = \mathbf{R}, \mathbf{C}, \mathbf{H})$ denote the K-module of dimension n and define a K-valued bilinear form on V by $\langle v, w \rangle = {}^t v\bar{w}, v, w \in V$. Then $O(n; \mathbf{R}), U(n), Sp(n)$ correspond to the linear "isometry" groups of V with $K = \mathbf{R}, \mathbf{C}, \mathbf{H}$ respectively.

v) $SO(n; \mathbf{R}) < O(n; \mathbf{R}), SU(n) < U(n)$ are the orientation preserving (all of determinant 1) subgroups.

vi) $Spin(n)(n > 2)$ is defined to be the universal covering group of $SO(n; \mathbf{R})$ and as such its existence can be established topologically. However, we will give an algebraic construction following [A-B-S]: Let CL_n denote the Clifford algebra associated with $\mathbf{R}^n$. It is generated as a $\mathbf{R}$-algebra by the basis $\varepsilon_1, \ldots, \varepsilon_n$ in $\mathbf{R}^n$ and the relations $\varepsilon_i\varepsilon_j + \varepsilon_j\varepsilon_i = 0(i \neq j), \varepsilon_i^2 = -1$. Consider the subgroup $Pin(n)$ of CL_n generated multiplicatively by $S^{n-1} = \{e = x^i\varepsilon_i : \sum(x^i)^2 = 1\}$. There is the projection $\pi : Pin(n) \to O(n; \mathbf{R})$ determined by $\pi(e) =$ the reflection about the hyperplane orthogonal to $e, e \in S^{n-1}$. Since $O(n; \mathbf{R})$ is generated by reflections, π is surjective and $\ker \pi = \{\pm I\}$. Put $Spin(n) = \pi^{-1}SO(n; \mathbf{R})$. We record that $Spin(n)$ is a double cover of $SO(n; \mathbf{R})$ and that $\pi_1(SO(n; \mathbf{R})) = \mathbf{Z}_2$.

vii) $T^n = S^1 \times \ldots \times S^1$ (n-torus) is an abelian Lie group. In fact, as we shall see any connected compact abelian Lie group is a torus. Note that $T^n \cong \mathbf{R}^n/\mathbf{Z}^n$.

viii) $Aut(\mathbf{O})$ = the algebra automorphisms of $\mathbf{O}$ is a connected simply connected compact 14 dimensional Lie group, usually denoted by G_2.

Observations.

i) $SO(2;\mathbf{R}) = \left\{ \begin{pmatrix} \cos 2\pi\theta & -\sin 2\pi\theta \\ \sin 2\pi\theta & \cos 2\pi\theta \end{pmatrix} : 0 \le \theta < 1 \right\} \cong \mathbf{R}/\mathbf{Z} = S^1 \cong$

$U(1) = \{ e^{2\pi i\theta} : 0 \le \theta < 1 \}$.

ii) $SU(2) = \left\{ \begin{pmatrix} z & w \\ -\bar{w} & \bar{z} \end{pmatrix} : z, w \in \mathbf{C}, |z|^2 + |w|^2 = 1 \right\} \cong S^3$. Indeed, it can be shown that S^1 and S^3 are the only spheres that are also Lie groups.

iii) $Sp(1) = \{ x \in \mathbf{H} : x\bar{x} = 1 \} \cong S^3$.

iv) $S^3 \cong Spin(3)$. To see this note that $SO(3;\mathbf{R})$ is homeomorphic to RP^3: a rotation is determined by an axis of rotation and an angle of rotation $-\pi \le \theta \le \pi$. Thus $SO(3;\mathbf{R})$ is homeomorphic to the closed ball of radius π in $\mathbf{R}$ with antipodal boundary points identified. Now the universal cover of RP^3 is S^3.

v) $Spin(4) \cong S^3 \times S^3, Spin(5) \cong Sp(2), Spin(6) \cong SU(4)$. These isomorphisms can be observed by looking at their Dynkin diagrams for example. See §6.

vi) There are the inclusions $O(n;\mathbf{R}) \hookrightarrow U(n) \hookrightarrow Sp(n)$.

vii) One has the injections

$$U(n) \to SO(2n;\mathbf{R}), \quad A + iB \mapsto \begin{pmatrix} A & -B \\ B & A \end{pmatrix}, \quad A, B \in M(n;\mathbf{R}),$$

$$Sp(n) \to SU(2n), \quad A + jB \mapsto \begin{pmatrix} A & -B \\ B & \bar{A} \end{pmatrix}, \quad A, B \in M(n;\mathbf{R}).$$

§2. Lie Algebra: Basic Results

In this section we discuss basic results concerning Lie algebras and Lie groups establishing a strong link between them. Let us begin with

Definition. A vector space ℓ over a field F with a multiplication (called the bracket operation) $\ell \times \ell \to \ell$, denoted by $(X, Y) \mapsto [X, Y]$, is called an (abstract) Lie algebra over F if

i) the bracket operation is bilinear,

ii) $[X, X] = 0$ for every $X \in \ell$,

iii) $[X, [Y, Z]] + [Y, [Z, X]] + [Z, [X, Y]] = 0$ for every $X, Y, Z \in \ell$.

From now on a Lie algebra will always be a Lie algebra over $\mathbf{R}$ unless otherwise specified.

We have the usual algebraic notions such as Lie algebra homomorphism, Lie subalgebra, ideal and so on.

If G is a Lie group then the set of left invariant vector fields in G forms a Lie algebra by letting $[X, Y]f = X(Yf) - Y(Xf)$ where f is a function on G. This Lie algebra, denoted by g, is a subalgebra of the infinite dimensional Lie algebra $\varkappa(G)$ (the set of all smooth vector fields in G) and is called the Lie algebra of G. Observe that $g \cong T_e(G)$ (= the tangent space of G at the identity).

Definition. A Lie group homomorphism $\theta \colon (\mathbf{R}, +) \to (G, \cdot)$ is called a one-parameter subgroup of G.

Proposition. There is a bijective correspondence between the set of one-parameter subgroups of G and $g \cong T_e(G)$.

Proof. We will show that the assignment $\theta \mapsto \theta_{*0}(1)$ is bijective.

$$
\begin{array}{ccc}
T_0(\mathbf{R}) = \mathbf{R} & \xrightarrow{\;\theta_{*0}\;} & T_e(G) = g \\
\Big\downarrow{\scriptstyle \mathrm{id}} & & \Big\downarrow \\
\mathbf{R} & \xrightarrow[\;\theta\;]{} & G
\end{array}
$$

Let θ be given. Consider the action of $\mathbf{R}$ on G given by $\Theta \colon \mathbf{R} \times G \to G, (t, g) \mapsto \mathbf{R}_{\theta(t)} g = g \cdot \theta(t)$. For $a \in G, L_a \Theta(t, g) = a \cdot g \cdot \theta(t) = \Theta(t, L_a(g))$. So Θ is left invariant. Thus its infinitesimal generating vector field V is left invariant. We have $V_e = \Theta(0, e) = \theta(0) = \theta_{*0}(1)$. We now let $X \in g$. Since X is left invariant it is complete. Thus there exists a unique infinitely extendable integral curve through any $g \in g$. Call this curve $\Theta_g \colon \mathbf{R} \to G$. Put $\theta(t) = g^{-1} \Theta_g(t)$. q.e.d.

Definition. The exponential map $\exp : g \to G$ is defined to be the map that makes the above diagram commutative for every θ.

So for $X \in g$ $\exp(X) = \theta(1)$ where θ is the one-parameter subgroup corresponding to X.

Examples.

i) $\exp : g \cong \mathbf{R} \to \mathbf{R}, \quad X \mapsto X.$

ii) $\exp : g \cong \mathbf{R} \to S^1, \quad X \mapsto e^{2\pi i X}.$

iii) $\exp : g \cong M(n; \mathbf{R}) \to GL(n; \mathbf{R}), \quad X \mapsto \sum X^n/n!.$

Remarks.

i) $\exp$ is a local analytic diffeomorphism.

ii) $\exp(a + b)X = \exp aX \cdot \exp bX$ for $a, b \in \mathbf{R}.$

iii) $[X, Y] = 0$ implies that $\exp(X + Y) = \exp X \cdot \exp Y$. Otherwise $\exp(X + Y)$ is given by the Campbell-Baker-Hausdorff formula.

Proposition. A connected abelian Lie group is isomorphic to $S^1 \times \ldots \times S^1 \times \mathbf{R} \times \ldots \times \mathbf{R}.$

Proof. We make three observations about a connected abelian Lie group: 1) G abelian implies that $[X, Y] = 0$ for every $X, Y \in g$, hence $\exp(X+Y) = \exp X \cdot \exp Y$; 2) G connected implies that $G = \cup U^i (1 \leq i < \infty)$, where U is a neighborhood of $e \in G$; 3) Wlog, $U = \exp(U_0)$, where U_0 is a neighborhood of $0 \in g$ diffeomorphic to U. The above observations put together at once imply that $\exp$ is surjective. Hence $G = g/\ker(\exp)$. Now $\ker(\exp) \cap U_0 = \{0\}$. So $0 \in \ker(\exp)$ is open in $\ker(\exp)$. Since translations are homeomorphisms any $X \in \ker(\exp)$ is open in $\ker(\exp)$. Thus $\ker(\exp)$ is a discrete subgroup of $(g, +)$. It is well-known that such a group is isomorphic to $\mathbf{Z}^p$, p is a nonnegative integer. So $G \cong g/\mathbf{Z}^p \cong S^p \times \mathbf{R}^{n-p}$. q.e.d.

Definition.

i) Inn: $G \to \operatorname{Aut}(G), \quad \operatorname{Inn}_g : h \mapsto ghg^{-1}, h \in G.$

ii) Ad: $G \to GL(g), \quad g \mapsto \operatorname{Inn}_{g*e}.$

iii) ad: $g \to g\ell(g), \quad \operatorname{ad} = \operatorname{Ad}_{*e}.$

(The notation $g\ell(g)$ means the Lie algebra of $GL(g)$.)

Remark. Let ℓ be an abstract Lie algebra. We then define ad: $\ell \to g\ell(\ell)$ (= the endomorphisms of ℓ) by $\operatorname{ad}_X(Y) = [X, Y]$. It is not hard to show that if ℓ is the Lie algebra of a Lie group then the two definitions coincide.

The following diagrams are easily seen to be commutative.

$$
\begin{array}{ccc}
g & \xrightarrow{\mathrm{Ad}_g} & g \\
{\scriptstyle \exp}\downarrow & & \downarrow \\
G & \xrightarrow[\mathrm{Inn}_g]{} & G
\end{array}
\qquad\qquad
\begin{array}{ccc}
g & \xrightarrow{\mathrm{ad}} & g\ell(g) \\
{\scriptstyle \exp}\downarrow & & \downarrow{\scriptstyle \exp} \\
G & \xrightarrow[\mathrm{Ad}]{} & GL(g)
\end{array}
$$

Proposition. Let G be a Lie group. Then

i) $\ker(\mathrm{ad}) = \mathbf{Z}(g)$,

ii) for connected G, $\ker(\mathrm{Ad}) = \mathbf{Z}(G)$,

iii) the Lie algebra of $\mathbf{Z}(G)$ is $\mathbf{Z}(g)$.

Proof. i) $\mathrm{ad}_X = 0$ iff $[X,Y] = 0$ for every $Y \in g$. By definition $\mathbf{Z}(g) = \{X \in g : [X,Y] = 0$ for every $Y \in g\}$. ii) Now $\mathbf{Z}(G) = \{g \in G : gh = hg$ for every $h \in G\}$. Since G is assumed to be connected for any $g \in G$ there are $X_1, \ldots, X_n$ such that $g = \exp X_1 \ldots \exp X_n$: just note that $G = \cup U^i (1 \le i < \infty)$, where $U = \exp(U_0)$ with $\exp|_{U_0}$ a diffeomorphism. It follows that $g \in \mathbf{Z}(G)$ iff $\exp X = g(\exp X)g^{-1} = \exp \mathrm{Ad}_g X$. So $g \in \mathbf{Z}(G)$ iff $\exp tX = \exp \mathrm{Ad}_g(tX)$ for $t \in \mathbf{R}, X \in g$. This means $\exp$ and $\exp \circ \mathrm{Ad}_g$ agree on U_0. But $\exp$ is injective in U_0, hence $\mathrm{Ad}_g = I$, i.e., $g \in \ker(\mathrm{Ad})$. iii) This follows from the general fact that the Lie algebra of the kernel of a Lie group homomorphism is the kernel of the derivative at the identity. q.e.d.

Since the Lie algebra of a Lie group is identified with the tangent space at the identity it is pretty evident that two "locally isomorphic" Lie groups would yield isomorphic Lie algebras. An elementary application of homotopy theory gives

Proposition. Let G, G' be connected and simply connected Lie groups. Then G is isomorphic to G' if and only if g is isomorphic to g'.

Proposition. Let G be a Lie group with the Lie algebra g. Then there is a bijective correspondence between subalgebras of g and connected C^ω-immersed subgroups (called the analytic subgroups) of G under which ideals correspond to normal subgroups.

Proof. A subspace m of g defines a left invariant distribution on G. This distribution is completely integrable iff m is a subalgebra. Let m be a subalgebra. Then the maximal integral submanifold through $e \in G$ of the corresponding distribution is the analytic subgroup of m. The rest is easy. q.e.d.

Definition. Let ℓ be a Lie algebra.

i) $\ell^{(0)} = \ell, \ell^{(1)} = [\ell^{(0)}, \ell^{(0)}], \ldots, \ell^{(i)} = [\ell^{(i-1)}, \ell^{(i-1)}], \ldots$ are all ideals in ℓ and $\ell > \ell^{(1)} > \ldots$ is called the derived series. We say ℓ is solvable if $\ell^{(n)} = 0$ for some n.

ii) $\ell^0 = \ell, \ell^1 = [\ell, \ell^0], \ldots, \ell^i = [\ell, \ell^{i-1}], \ldots$ are all ideals in ℓ and $\ell > \ell^1 > \ldots$ is called the descending central series. We say ℓ is nilpotent if $\ell^n = 0$ for some n.

Clearly abelian implies nilpotent and nilpotent implies solvable. The standard examples of a nilpotent Lie algebra and a solvable Lie algebra are strictly upper triangular matrices and upper triangular matrices respectively.

Definition. Let ℓ be a Lie algebra.

i) ℓ is said to be simple if $\ell \neq \mathbf{R}$ and it does not contain a proper ideal.

ii) ℓ is said to be semisimple if its radical (= the maximal solvable ideal) is trivial.

Obviously, simple implies semisimple. Indeed, one can show that any semisimple Lie algebra is the direct sum of its simple ideals.

Theorem (Cartan). A Lie algebra ℓ is semisimple if and only if its Cartan-Killing form $\kappa(X, Y) = \operatorname{tr}(\operatorname{ad} X \operatorname{ad} Y) : \ell \times \ell \to \ell$ is nondegenerate.

Proof. See [Hu] p. 22.

A Lie group is called simple (semisimple) if its Lie algebra is simple (semisimple). So a Lie group is simple iff it is nonabelian and does not contain any proper connected normal subgroup.

Propositon. A connected semisimple Lie group is compact if and only if its Cartan-Killing form (on its Lie algebra) is negative definite.

Proof. See [He] p. 132.

Examples.

i) $GL(n; \mathbf{R})$ is not semisimple.

ii) $U(n)$ is not simple. $\mathbf{Z}(U(n)) = \{e^{it}I\} \cong S^1$ is a connected normal subgroup of $U(n)$.

iii) $SO(n; \mathbf{R})$ is semisimple for $n > 2$. $\kappa(X, Y) = (n-2)\mathrm{tr}(XY)$.

iv) $SO(n; \mathbf{R})(n = 3 \text{ or } n > 4), SU(n)(n > 1), Sp(n)$ are all simple.

In §6 we will give a classification of compact simple Lie groups up to Lie algebra.

§3. Orthogonal and Unitary Representations

Let G be a Lie group and also let V be a vector space over $K = \mathbf{R}$ or $\mathbf{C}$.

Definition. A representation of G on V is a homomorphism $\rho : G \to GL(V)$.

Upon choosing a basis in $V, \rho : G \to GL(n; K), n = \dim V$.

Notation. Whenever there is no possibility of confusion we write gv instead of writing $\rho(g)v, v \in V$.

Observation. The following conditions characterize the notion of representation: for every $g, g' \in G$ and $v \in V$,

i) $ev = v, g(g'v) = (gg')v$,

ii) gv is K-linear in v,

iii) gv is smooth in g and in v.

Not surprisingly, two representations $\rho, \rho' : G \to GL(V), GL(W)$ are said to be isomorphic if there exists an equivariant vector space isomorphism $f : V \to W$, i.e., $f(gv) = g(fv), v \in V$.

Constructions. Let $\rho, \rho' : G \to GL(V), GL(V')$ be representations. We then have the following induced representations.

i) $\rho \oplus \rho' : G \to GL(V \oplus V')$.

ii) $\rho^* : G \to GL(V^*), \rho^*(g)f(v) = f(\rho(g)^{-1}v), f \in V^*, v \in V$.

iii) $\rho \otimes \rho' : G \to GL(V \otimes V'), \rho \otimes \rho'(g)(v \otimes v') = \rho(g)v \otimes \rho'(g)v'$.

iv) $\mathrm{Hom}(V, V') = V^* \otimes V'$ and this gives $\rho^* \otimes \rho' : G \to \mathrm{Hom}(V, V')$.

v) $S^n V$ (symmetric product), $\Lambda^n V$ (skew-symmetric product) also inherit representations in an obvious manner.

The following theorem from analysis is basic in representation theory of compact Lie groups.

Theorem (Harr integral). Let G be a compact Lie group. Also let $C(G)$ denote the space of real valued continuous functions on G. Then there exists a unique map $\int_G : C(G) \to \mathbf{R}$ such that

i) $\int_G$ is a positive linear functional,

ii) $\int_G 1 = 1$,

iii) $\int_G$ is biinvariant, i.e., $\int_G f = \int_G L_g^* f = \int_G R_g^* f$, where L_g, R_g denote the left and right translations by $g \in G$.

As an application of the above theorem we prove a theorem on the existence of biinvariant metrices on certain Lie groups. (A biinvariant metric on a Lie group is a metric invariant under left and right translations.) Let $\langle , \rangle$ denote a left invariant metric on G. Then $\langle , \rangle$ is also right invariant iff $\langle , \rangle_e$ gives an $\mathrm{Ad}(G)$-invariant inner product in $g \cong T_e G$. (Easy exercise.) This enables us to identify biinvariant metrics on G with $\mathrm{Ad}(G)$-invariant inner product in g.

Theorem. There exists a biinvariant metric on a Lie group G if and only if the closure of $\mathrm{Ad}(G)$ in $GL(g)$ is compact.

Proof. Let $\langle , \rangle$ denote a biinvariant metric on G. Then $\langle , \rangle_e$ is $\mathrm{Ad}(G)$-invariant. This means that $\mathrm{Ad}(G)$ lies in $O(g, \langle , \rangle_e) = O(n) = $ the orthogonal group. But $O(n)$ is compact, so $\mathrm{cl}(\mathrm{Ad}(G))$ is also compact. Conversely, assume that $H = \mathrm{cl}(\mathrm{Ad}(G))$ in $GL(g)$ is a compact Lie group. Let $\int_H = \int_H \cdot d\mu(h) : C(H) \to \mathbf{R}$ denote the Harr integral of H. Also let $\langle , \rangle_e'$ be any inner product in $g(\cong \mathbf{R}^n)$. Define $\langle X, Y \rangle_e = \int_H \langle hX, hY \rangle_e' d\mu(h)$, $X, Y \in g$. Now for $x \in H$, $\langle xX, xY \rangle_e = \int_H \langle hxX, hxY \rangle_e' d\mu(h) = \int_H R_x^* f(h) d\mu(h)$ (here $f : H \to \mathbf{R}$ given by $f(h) = \langle hX, hY \rangle_e') = \int_H f(h) d\mu(h) = \langle X, Y \rangle_e$.

In particular, $\langle X, Y \rangle_e = \langle hX, hY \rangle_e$ for every $h \in \mathrm{Ad}(G)$. Thus $\langle X, Y \rangle_e = \langle \mathrm{Ad}_g X, \mathrm{Ad}_g Y \rangle$ for every $g \in G$. q.e.d.

Consequently there are biinvariant metrics on compact Lie groups.

Remark. We mention that for a connected compact simple Lie group any two biinvariant metrics are constant multiples of each other.

A slight modification of the above proof yields

Theorem. Let G be a compact Lie group and $\rho : G \to GL(V)$, a (real or complex) representation. Then there exists a ρ-invariant (real or hermitian) inner product in V.

Corollary. Let G be a connected compact Lie group and $\rho : G \to GL(V)$, a one-dimensional real representation, i.e., $V \cong \mathbf{R}$. Then ρ is trivial.

Proof. On V impose a ρ-invariant inner product. Then with respect to an orthonormal basis of this inner product we can regard $\rho : G \to GL(V)$ as taking values in $O(1; \mathbf{R}) = \{\pm 1\}$. Since G is connected $\rho(G) = \{1\}$. q.e.d.

Definition. A (real or complex) representation $\rho : G \to GL(V)$ is said to be irreducible if there does not exist a proper invariant subspace W of V ($\rho(G)W < W$).

Example. Any Lie group G acts on its Lie algebra by Ad: $G \to GL(\mathfrak{g})$. Then G is simple iff Ad is irreducible and dim $G > 1$.

Proposition. Let G be a compact Lie group and $\rho : G \to GL(V)$ any representation (whether real or complex). Then ρ is the direct sum of irreducible representations on its invariant subspaces.

Proof. Let $W < V$ be a proper invariant subspace of V. (If such a subspace does not exist then ρ is irreducible and we are done.) Choose a ρ-invariant inner product in V and let $U = $ the orthogonal (or unitary) complement of W in V. Then U is also an invariant subspace. The rest follows by induction on the dimension of V. q.e.d.

Proposition. Let G be a connected abelian group. Then every irreducible complex representation $\rho : G \to GL(V)$ is one-dimensional, $V \cong \mathbf{C}$.

Proof. We first state Schur's Lemma without proof: If a linear map $f : V \to V$ commutes with $\rho(g)$ for every $g \in G$, where ρ is an irreducible complex representation then $f(v) = \lambda v$ for some $\lambda \in \mathbf{C}$, $v \in V$. Now put $f_h = \rho(h) \in GL(V)$. Then G abelian implies that

$\rho(g) \circ f_h = f_h \circ \rho(g)$ for every $g \in G$. By Schur's Lemma f_h then is a scalar multiplication. But this implies that any subspace of V is ρ-invariant. The result now easily follows. q.e.d.

In the above proposition if we further assume G to be compact then we can write $\rho : G \to U(1)$.

Note that $G \cong T^n$ from §2. It is not hard to show that a homomorphism $T^n \to T^1 \cong \mathbf{R}/\mathbf{Z}$ has the form $(x_1, \ldots, x_n) \mapsto \Sigma m_i x_i \pmod{\mathbf{Z}}$ for some $m_1, \ldots, m_n \in \mathbf{Z}$. It follows that any irreducible unitary representation $\rho : T^n \to U(1)$ is of the form $(x_1, \ldots, x_n) \mapsto e^{2\pi i \Sigma m_i x_i}$. Realification yields

Proposition. Any irreducible orthogonal representation of T^n is of the form

i) $\rho : T^n \to SO(1; \mathbf{R})$ trivial, or

ii) $\rho : T^n \to SO(2; \mathbf{R}) \cong U(1)$ given by

$$(x_1, \ldots, x_n) \mapsto \begin{pmatrix} \cos 2\pi \sum m_i x_i & -\sin 2\pi \sum m_i x_i \\ \sin 2\pi \sum m_i x_i & \cos 2\pi \sum m_i x_i \end{pmatrix},$$

where (m_i) are integers not all zero.

§4. Maximal Tori and Stiefel Diagrams

From now on our working assumption will be that a Lie group is connected and compact. (When stating a theorem we will make the hypotheses explicit.) The reasons for this are twofold: many results apply only to connected compact groups (the compactness assumption being quite essential.) In addition, understanding of the compact case is often sufficient; for example, this certainly is the case in studying symmetric spaces by virtue of duality.

Definition. Let G be a Lie group and also let $H < G$, a subgroup.
   i) H is called a torus in G if it is closed and if $H \cong T^n$ for some positive integer n.
   ii) H is called a maximal torus in G if it is a torus in G and for any torus H' in G containing H we have $H' = H$.
Any torus in G is clearly contained in a maximal torus.

Notation. T will always denote a maximal torus and S, a torus in G.

Lemma. Let G be a compact Lie group and also let H be a connected abelian subgroup of G. Then the closure of H in G is also a connected abelian subgroup. (Hence it is a torus in G.)

Proof. A simple continuity argument does the trick. q.e.d.

Consequently, if G is compact then maximal tori are automatically closed in G.

Corollary. S is a maximal torus in a compact Lie group G if and only if S is a maximal connected abelian subgroup of G.

Recall that a subspace $m < g$ is said to be abelian if $[X, Y] = 0$ for every $X, Y \in m$. So an abelian subspace is a subalgebra. $\exp(m)$ in a compact group is then a torus if it is topologically closed. Conversely the Lie algebra of a torus is an abelian subalgebra of g.

Definition. Let G be a compact Lie group. Then a subset m of g is called a Cartan subalgebra if it is a maximal abelian subalgebra.

Proposition. Let G be a compact Lie group. Then a torus S is maximal if and only if its Lie algebra is a Cartan subalgebra.

Proof. Let S be a maximal torus. Then certainly its Lie algebra $\mathfrak{s}$ is an abelian subalgebra of g. By dimension argument we see that $\mathfrak{s}$ is also maximal. Conversely, let S be a torus in G and assume that $\mathfrak{s}$ is a maximal abelian subalgebra of g. S being a torus $\exp(\mathfrak{s}) = S$. Now the maximality of $\mathfrak{s}$ implies that S is a maximal torus. q.e.d.

The following theorem by Cartan is basic.

Theorem (Cartan). Let G be a connected compact Lie group. Fix a maximal torus T in G and let t denote its Lie algebra. Then

i) $G = \cup \operatorname{Inn}_g T, g \in G$,

ii) $g = \cup \operatorname{Ad}_g t, g \in G$,

iii) if T' is any other maximal torus in G then for some $g \in G, T' = \operatorname{Inn}_g T$, and $t' = \operatorname{Ad}_g t$.

Proof. See [B-D] p. 159.

It follows that any two maximal tori in a connected compact Lie group G are of the same dimension, and this dimension is called the rank of G. Another immediate consequence of the above theorem is that for a connected compact Lie group the exponential map is surjective.

Let S be a torus in a connected compact Lie group G. We consider the adjoint representation of G restricted to S, $\text{Ad}|_S : S \to GL(\ )$. Impose a biinvariant metric on G, which exists since G is compact. This gives an $\text{Ad}(G)$-invariant inner product in g and relative to an orthonormal basis of this inner product we have $\text{Ad}(S) < SO(m; \mathbf{R})$, where $m = \dim G$. Such a representation decomposes into the direct sum of irreducible representations on its invariant subspaces with the dimensions of invariant subspaces 1 or 2. Write $\text{Ad}|_S = \rho_0 \oplus \sum \rho_i = \text{Ad}|_{V_0} \oplus \sum \text{Ad}|_{V_i} (i > 0)$, where $\dim V_i = 2$ and $V_0 =$ the direct sum of one-dimensional invariant subspaces. Observe that $V_0 =$ the Lie algebra of S iff $S = T$, a maximal torus. We let $S = T$ and also let $\{E_\alpha\}$ denote an orthonormal basis for g so that E_{2i-1}, E_{2i} span V_i for $i > 0$. Thus with respect to E_{2i-1}, E_{2i} we have $\rho_i : T \to SO(2; \mathbf{R})$,

$$t \mapsto \begin{pmatrix} \cos 2\pi\Theta_i(t) & -\sin 2\pi\Theta_i(t) \\ \sin 2\pi\Theta_i(t) & \cos 2\pi\Theta_i(t) \end{pmatrix} = e^{2\pi i \Theta_i(t)}$$

(using the identification $SO(2) = U(1)$.)

Notation. Throughout this section we will adhere to: $r = \dim T = \text{rank}\, G$, $2n = \dim G - \dim T$.

Definition. Let G be a connected compact Lie group and T, a maximal torus in G as in the preceding discussion. Then $\Delta = \{\pm\Theta_i : T \to \mathbf{R}/\mathbf{Z} = S^1\}$ are called the roots of G relative to T. $\{\pm\Theta_{i*e} = \pm\theta_i\}$ are also called the roots of G. We will confuse the two and let the context dictate the meaning.

For obvious geometrical reason $\exp^{-1}\{e\} \cap t$ is called the integer lattice of t. Note that θ_i takes integer values on the integer lattice.

$$
\begin{array}{ccc}
\mathbf{R}^r \cong t & \xrightarrow{\ \theta_i\ } & \mathbf{R} \\
\downarrow{\scriptstyle \exp} & & \downarrow{\scriptstyle \exp} \\
T & \xrightarrow[\ \Theta_i\ ]{} & S^1
\end{array}
$$

ker Θ_i is a closed codimension 1 subgroup of T, and the Lie algebra of ker Θ_i is ker θ_i.

Proposition. $Z(g) = \cap \ker \Theta_i, 1 \le i \le n$.

Proof. Notice that $Z(G) < Z(T) = T$. Now $\mathrm{Inn}_t \circ \exp = \exp \circ \mathrm{Ad}_t$ and exp is surjective. It follows that $\mathrm{Ad}_t = \mathrm{id}$. So $t \in \ker \Theta_i$ iff $\mathrm{Inn}_t = \mathrm{id}$ iff $t \in Z(G) \cap T = Z(G)$. q.e.d.

Definition. Let G be a connected compact Lie group and T, a fixed maximal torus in G. The Stiefel diagram of G is the vector space $t = \mathbf{R}^r$ with the hyperplanes $\exp^{-1}\{\ker \Theta_i\}, 1 \le i \le n$. The infinitesimal Stiefel diagram is the vector space $= \mathbf{R}^r$ with the hyperplanes ker $\theta_i, 1 \le i \le n$.

Examples.

i) $G = U(n)$.
$$T = \{\mathrm{diag}(e^{2\pi i x_1}, \ldots, e^{2\pi i x_n}) = t : x_i \in \mathbf{R}/\mathbf{Z} = S^1\},$$
$$t = \{\mathrm{diag}(id_1, \ldots, id_n) : d_i \in \mathbf{R}\}.$$
Let $E_{ij} =$ the $n \times n$ matrix with $+1$ at (i, j)-entry, -1 at (j, i)-entry and zeros elsewhere. Also let $F_{ij} =$ the $n \times n$ matrix $+1$ at (i, j) and (j, i)-entries and zeros elsewhere. Then $g = t \oplus \Sigma R E_{ij} \oplus \sqrt{-1} R F_{ij}, i < j$. Let $X_{ij} = x E_{ij} + \sqrt{-1} y F_{ij}, x, y \in \mathbf{R}$. Then $\mathrm{Ad}_t X_{ij} = t X_{ij} t^{-1} =$ the $n \times n$ skew hermitian matrix with nonzero entries only at (i, j) and (j, i) spots with $e^{2\pi i(x_i - x_j)}(x + iy)$ at (i, j)-entry. It follows that $\Theta_{ij} : T \to S^1$ is given by $t \mapsto x_i - x_j (\mathrm{mod}\, \mathbf{Z})$ and $\theta_{ij} : = \mathbf{R}^n = \{x = (x_1, \ldots, x_n)\} \to \mathbf{R}$ by $\theta_{ij}(x) = x_i - x_j$. Thus $\Delta = \{\pm(x_i - x_j) : 1 \le i < j \le n\}$. (It is understood that by $x_i - x_j$ it is meant the function $\mathbf{R}^n \to \mathbf{R}$ given by $(x_1, \ldots, x_n) \mapsto x_i - x_j$.) $|\Delta| = n(n-1)$, rank $G = n, \dim G = n + n(n-1)$.

ii) $G = SU(n)$.
T as in i) except that we require $\Sigma x_i = 0 (\mathrm{mod}\, \mathbf{Z})$.
t as in i) except that we require $\Sigma d_i = 0$.
Going through the same computation as in i) we get $\Delta = \{\pm(x_i - x_j) : 1 \le i << \le n\}$.

iii) $G = Sp(n)$.
$T, t, E_{ij}, F_{ij}, X_{ij}$ as in i).

Then $g = t \oplus \Sigma RE_{ij} \oplus iRF_{ij}(i < j) \oplus \Sigma jRF_{ij} \oplus kRF_{ij}(i \leq j)$. Note that i, j, k other than indices are quarternions. Put $Y_{ij} = (x + iy)jF_{ij}$. Then $\mathrm{Ad}_t Y_{ij} = e^{2\pi i(x_i + x_j)} Y_{ij}$. It follows that $\Delta = \{\pm(x_i - x_j), \pm(x_i + x_j), \pm 2x_i, 1 \leq i < j \leq n\}$.

iv) $G = SO(2n; \mathbf{R})$.

Recall that $U(n) \hookrightarrow SO(2n; \mathbf{R})$ via the standard injection.

$$T = \left\{ \mathrm{diag}(D_1, \dots, D_n) : D_i = \begin{pmatrix} \cos 2\pi x_i & -\sin 2\pi x_i \\ \sin 2\pi x_i & \cos 2\pi x_i \end{pmatrix}, \right.$$
$$\left. x_i \in \mathbf{R}/\mathbf{Z} = S^1 \right\},$$

$$t = \left\{ \mathrm{diag}\left(\begin{pmatrix} 0 & -d_1 \\ d_1 & 0 \end{pmatrix}, \dots, \begin{pmatrix} 0 & -d_n \\ d_n & 0 \end{pmatrix} \right) : d_i \in \mathbf{R} \right\}.$$

$g = t \oplus \Sigma RE_{ij} \oplus RE_{ij} \oplus RE'_{ij} \oplus RE'_{ij}(1 \leq i < j \leq n)$:

$E_{ij} =$ the $2n \times 2n$ matrix with $+1$ at $(2i - 1, 2j - 1)$ and $(2i, 2j)$-entries, -1 at $(2j - 1, 2i - 1)$ and $(2j, 2i)$-entries and zeros elsewhere; $F_{ij} =$ the $2n \times 2n$ matrix $+1$ at $(2i, 2j - 1)$ and $(2j, 2i - 1)$-entries, -1 at $(2i - 1, 2j)$ and $(2j - 1, 2i)$-entries and zeros elsewhere; $E'_{ij} =$ the $2n \times 2n$ matrix $+1$ at $(2i - 1, 2j - 1)$ and $(2j, 2i)$-entries, -1 at $(2i, 2j)$ and $(2j - 1, 2i - 1)$-entries and zeros elsewhere; $F'_{ij} =$ the $2n \times 2n$ matrix $+1$ at $(2i - 1, 2j)$ and $(2i, 2j - 1)$-entries, -1 at $(2j - 1, 2i)$ and $(2j, 2i - 1)$-entries and zeros elsewhere. Put $t = \mathrm{diag}(D_1, \dots, D_n) \in T$. Also put $v = xE_{ij} + yF_{ij}, v' = xE'_{ij} + yF'_{ij}$. Then

$$\mathrm{Ad}_t : v = \begin{pmatrix} x \\ y \end{pmatrix} \mapsto \begin{pmatrix} \cos 2\pi(x_i - x_j) & -\sin 2\pi(x_i - x_j) \\ \sin 2\pi(x_i - x_j) & \cos 2\pi(x_i - x_j) \end{pmatrix} \begin{pmatrix} x \\ y \end{pmatrix} \text{ and}$$

$$v' = \begin{pmatrix} x \\ y \end{pmatrix} \mapsto \begin{pmatrix} \cos 2\pi(x_i + x_j) & -\sin 2\pi(x_i + x_j) \\ \sin 2\pi(x_i + x_j) & \cos 2\pi(x_i + x_j) \end{pmatrix} \begin{pmatrix} x \\ y \end{pmatrix},$$

where v and v' are written with respect to the basis E_{ij}, F_{ij} and E'_{ij}, F'_{ij} respectively. This gives the roots $\theta_{ij} = x_i - x_j, \theta'_{ij} = x_i + x_j$. So $\Delta = \{\pm(x_i - x_j), \pm(x_i + x_j), 1 \leq i < j \leq n\}$.

v) $G = SO(2n + 1; \mathbf{R})$.

We inject $SO(2n; \mathbf{R})$ into $SO(2n + 1; \mathbf{R})$ by $g \mapsto \begin{pmatrix} g & 0 \\ 0 & 1 \end{pmatrix}$. Then T of the previous example upon the injection is a maximal torus in G.

$g = O(2n; \mathbf{R}) \oplus \Sigma V_i \, (1 \leq i \leq n)$, where

$$V_i = \left\{ X_i = \begin{bmatrix} & & x \\ 0 & & i \\ & & y \\ -x, -y & & 0 \end{bmatrix} : x, y \in \mathbf{R} \right\}.$$

On X_i, t acts by rotation through the x_i-axis. This gives $\Delta = \{\pm x_i, x_i - x_j, \pm(x_i + x_j), i \neq j\}$.

Computation of centers. Recall that $\mathbf{Z}(G) = \cap \ker \Theta_i$.

i) $G = U(n)$: We must have $x_1 = x_2 = \ldots = x_n$ in the description of T in i) of the preceding example. Hence $\mathbf{Z}(G) = \{e^{2\pi i x} I : x \in \mathbf{R}/\mathbf{Z}\} \cong S^1$.

ii) $G = SU(n)$: In addition to $x_1 = x_2 = \ldots = x_n$ we also need $\Sigma x_i = 0 (\mathrm{mod}\, \mathbf{Z})$. Thus $\mathbf{Z}(G) = \{\omega I : \omega^n = 1\} \cong \mathbf{Z}_n$.

iii) $G = Sp(n)$: We need $x_i + x_j = 0(\mathrm{mod}\, \mathbf{Z})$ for every i, j. Hence $x_i = 0(\mathrm{mod}\, \mathbf{Z})$ for every i or $x_i = \frac{1}{2}(\mathrm{mod}\, \mathbf{Z})$ for every i. So $\mathbf{Z}(G) = \{\pm I\}$.

iv) $\mathbf{Z}(SO(2n; \mathbf{R})) = \{\pm I\}(n > 1)$, $\mathbf{Z}(SO(2; \mathbf{R})) = SO(2; \mathbf{R})$.

v) $\mathbf{Z}(SO(2n + 1; \mathbf{R})) = \{I\}$.

Examples of Stiefel Diagrams

i) $G = U(2)$. (Integer lattice points appear as circles.)

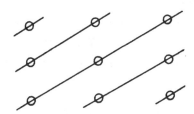

ii) $G = SO(4; \mathbf{R})$.

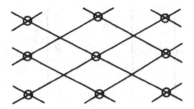

iii) $G = SO(5; \mathbf{R})$.

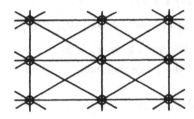

iv) $G = Sp(2)$.

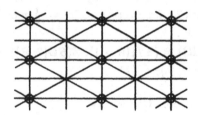

§5. Weyl Groups

Definition. Let G be a connected compact Lie group. Fix a maximal torus T in G. The Weyl group $W(G, T)$ is the group of automorphisms of T which are the restrictions of inner automorphisms of G.

Observations.

i) If $\phi \in W(G, T)$ then $\phi : t \mapsto gtg^{-1}$ for some $g \in N(T) =$ the normalizer of T in G. So $W(G, T)$ is naturally isomorphic to $N(T)/T$.

ii) $g \in N(T)$ iff $\operatorname{Inn}_g T$ iff $\operatorname{Ad}_g t = t$. So $W(G, T)$ acts on t as well.

iii) $W(G, T)$ acts on t^* via $(\phi \cdot f)(x) = f(\phi^{-1}(x))$, $x \in t, f \in t^*$.

iv) Any other maximal torus T' in G is conjugate to T by Cartan's theorem. Hence $W(G, T) \cong W(G, T')$ and the ensuing dependence on T is slight.

Fix a maximal torus in G and just write $W(G)$ from now on.

Proposition. $W(G)$ is a finite group.

Proof. Let N_e denote the identity component of $N(T)$. Since the full group of automorphisms of $T(\cong GL(r; \mathbb{Z}))$ is discrete we see thay N_e acts trivially on T, i.e., $ntn^{-1} = t$ for every $t \in T, n \in N_e$. Let H be an arbitrary one-parameter subgroup of N_e. Then the product $H \cdot T$ is a connected abelian group containing T, hence by maximality we must have $H \cdot T = T$. Since the elements of one-parameter subgroups of N_e generate N_e this implies that $N_e = T$, hence $W(G) = N(T)/N_e$. But $N(T)$ is compact so that $N(T)/N_e$ is compact and discrete. q.e.d.

Proposition. $W(G)$ permutes the roots $\Delta = \{\pm\theta_i\}$.

Proof. Let $\phi \in W(G)$ acts on T. Then it is not hard to see that the representations $\text{Ad}|_T$ and $\text{Ad}|_T \circ \phi$ are isomorphic. The rest follows from the general fact that isomorphic representations share the same set of roots. q.e.d.

Notation. $C(X) = $ the centralizer of a subset X of $G = \{g \in G : gx = xg$ for every $x \in X\}$. $N(X) = $ the normalizer of a subset X of $G = \{g \in G : gXg^{-1} = X\}$. Note that $C(x) = N(x), x \in G$.

Proposition. (G as in the above.)

i) For any $g \in G, C_e(g) = \cup T'$, where the union is taken over all maximal tori T' containing g.

ii) If S is a torus in G then $C(S) = \cup T'$, where the union is taken over all maximal tori T' containing S.

Proof. See [A] p. 97 or [B-D] p. 165.

Definition. Let G be a connected compact Lie group. Then

i) $g \in G$ is said to be regular if it is contained in just one maximal torus,

ii) $g \in G$ is said to be singular otherwise, i.e., it is contained in more than one maximal torus.

In view of the proposition immediately preceding the above definition we have

Observation.

i) $g \in G$ is regular iff $\dim C(g) = \operatorname{rank} G$,

ii) $g \in G$ is singular iff $\dim C(g) > \operatorname{rank} G$.

Proposition. $\cup \ker \Theta_i = \{\text{singular points in } T\}$.

Proof. Recall that $\operatorname{Ad}|_T = \rho_0 \oplus \Sigma \rho_i$ and that

$$\rho_i(t) = \begin{pmatrix} \cos 2\pi\Theta_i(t) & -\sin 2\pi\Theta_i(t) \\ \sin 2\pi\Theta_i(t) & \cos 2\pi\Theta_i(t) \end{pmatrix}.$$

So $t \in \ker \theta_i$ means that there exists a subspace of $t^\perp < g$ on which t acts trivially. Exponentiating, one obtains a one-parameter subgroup of G not contained in T fixed by Inn_t. q.e.d.

Geometrically speaking, the inverse image of singular points in T under exp consists of the hyperplanes of the Stiefel diagram.

Let G, T be as in the above and also let $\Delta = \{\pm\theta_i\}$ (in t^*) denote the roots of G. Then for $i \neq j$ we have $\theta_i \wedge \theta_j \neq 0$. That is to say, the roots are pairwise independent.

Definition. (Maintain the above notation.)

For each i pick $\varepsilon_i = \pm 1$. Then the set $\{x \in t : \varepsilon_i\theta_i(x) > 0$ for every $i\}$ is either empty or a nonempty convex set. In the latter case it is called a Weyl chamber (WC) and its closure is given by $\{x \in t : \varepsilon_i\theta_i(x) \geq 0$ for every $i\}$. (Thus the hyperplanes of the infinitesimal Stiefel diagram divide t into Weyl chambers.) A wall of a Weyl chamber is the intersection of its closure with a hyperplane $\ker \theta_i$ when this intersection has dimension equal to rank $G - 1$. The roots $\{\varepsilon_i\theta_i\}$ are called the positive roots and the corresponding Weyl chamber is called the fundamental Weyl chamber (FWC). We write $\Delta = \Delta_+ \cup \Delta_-$, where $\Delta_+ = \{\varepsilon_i\theta_i\}$ and $\Delta_- = \{-\varepsilon_i\theta_i\}$.

Since $W(G)$ permutes the roots it also permutes the hyperplanes of the infinitesimal Stiefel diagram. It follows that $W(G)$ acts on the set of Weyl chambers.

Theorem. $(G, T$ as in the above.)

$W(G)$ permutes the Weyl chambers simply transitively.

Proof. See [B-D] pp. 193–194.

Examples. (Review the examples in §4 for the notation.)

i) $G = U(n)$.

$g =$ skew hermitian $n \times n$ matrices. Define an $\mathrm{Ad}(G)$-invariant inner product by $\langle X, Y \rangle = \mathrm{tr}({}^t\overline{X}Y) = \mathrm{tr}(-XY) = -(1/2n)$ $\kappa(X, Y)$. So for $X = \mathrm{diag}(ix_1, \ldots, ix_n) \in t$ we get $\langle X, X \rangle = \Sigma x_i^2$. Upon polarization we get an inner product in . Take the root $\theta_{ij} = x_i - x_j$. This gives the hyperplane $x_i = x_j$ for ker θ_{ij} and the reflection ϕ_{ij} in this hyperplane is given by $(x_1, \ldots, x_n) \mapsto (x_1, \ldots, x_{i-1}, x_j, x_{i+1}, \ldots, x_{j-1}, x_i, x_{j+1}, \ldots, x_n)$. This θ_{ij} is induced by an inner automorphism Inn_g with

$$
g = \begin{bmatrix} I & & & \\ & 1 & & \\ & & -1 & \\ & & & I \end{bmatrix} \begin{matrix} \\ i \\ j \end{matrix} \text{, or any element in } gT.
$$

So $\phi_{ij} \in W(G)$. We see easily that these reflections generate $W(G)$. It follows that $W(G) = S(x_1, \ldots, x_n) \cong S_n$ (the symmetric group on n letters) and $|W(G)| = n!$.

ii) $G = SU(n)$.

Exactly the same calculation shows that $W(G) \cong S_n$.

iii) $G = Sp(n)$.

$W(G)$ consists of the maps

$$
(x_1, \ldots, x_n) \mapsto (\varepsilon_1 x_{\sigma(1)}, \ldots, \varepsilon_n x_{\sigma(n)}),
$$

where $\varepsilon_i = \pm 1, \sigma \in S_n$. So $|W(G)| = 2^n n!$.

iv) $G = SO(2n + 1; \mathbf{R})$.

Here, $W(G) \cong W(Sp(n))$.

v) $G = SO(2n; \mathbf{R})$.

$W(G)$ consists of the maps

$$
(x_1, \ldots, x_n) \mapsto (\varepsilon_1 x_{\sigma(1)}, \ldots, \varepsilon_n x_{\sigma(n)}),
$$

where $\varepsilon_i = \pm 1, \Pi \, \varepsilon_i = 1$, So $|W(G)| = 2^{n-1} n!$.

Remark. Let ϕ denote that reflection (relative to an $\mathrm{Ad}(G)$-invariant inner product) about $\ker \theta$ $(\theta \in \Delta)$ in t. Then we always have $\phi \in W(G)$ and it is not hard to show that such reflections actually generate $W(G)$.

§6. Dynkin Diagrams and the Classification

Let G be a connected compact Lie group and also let T denote a fixed maximal torus in G. Impose a biinvariant metric on G. This gives an $\mathrm{Ad}(G)$-invariant inner product in g and by restriction t becomes an inner product space as well. Note that $W(G) < O(t)$ with this inner product. Identify t with t^* via $x \mapsto i_x, i_x(y) = \langle x, y \rangle, x, y \in t$. t^* is then naturally an inner product space with $\langle i_x, i_y \rangle = \langle x, y \rangle$.

Let $\phi_i \in W(G)$ be the reflection in the hyperplane $\ker \theta_i$. We will let ϕ_i act on t^* using the above identification. So $\ker \theta_i$ is viewed as the hyperplane in t^* orthogonal to θ_i. $\phi_i \in GL(t^*)$ is explicitly given by

$$f \mapsto f - 2\langle \theta_i, f \rangle \theta_i / \langle \theta_i, \theta_i \rangle, \quad f \in t^*.$$

Definition. A weight of G (relative to T) is an element of t^* which takes integer values on the integer lattice.

So a root is a weight.

Proposition. Let f be a weight of G. Then $\phi_i(f) - f = n\theta_i$ for some $n \in \mathbf{Z}$.

Proof. Elementary vector arithmetic shows that $\phi_i(f) - f$ is a multiple of θ_i. But $\phi_i(f) - f$ is easily seen to be a weight. Thus it has to be an integral multiple of θ_i. q.e.d.

It follows at once that $2\langle \theta_i, f \rangle / \langle \theta_i, \theta_i \rangle$ is an integer for $\theta_i \in \Delta, f$ any weight.

Notation. $A_{ij} = 2\langle \theta_i, \theta_j \rangle / \langle \theta_i, \theta_i \rangle$, $\theta_i, \theta_j \in \Delta$.

Theorem. $(G, T$ as in the above.)

Let θ_i, θ_j be roots of G such that $\theta_i \neq \pm \theta_j$. Then there are only four possibilities for the (unoriented, modulo π) angle between θ_i and θ_j, namely, $90°, 60°, 45°, 30°$. Moreover,

i) if the angle is 60° then $|\theta_i| = |\theta_j|$,

ii) if the angle is 45° then the ratio of the lengths of two roots is $\sqrt{2}$,

iii) if the angle is 30° then the ratio is $\sqrt{3}$, and

iv) $\theta_j + k\theta_i$ is also a root where k is any integer between 0 and $-A_{ij}$. (These roots form so called the θ_i-root string through θ_j.)

Proof. Let $\measuredangle$ denote the unoriented modulo π angle between θ_i and θ_j.

Now $A_{ij} = 2\frac{|\theta_i|}{|\theta_j|}\cos\measuredangle$. It follows that $A_{ij} \cdot A_{ji} = 4\cos^2\measuredangle$. Thus $4\cos^2\measuredangle = 0, 1, 2, 3, 4$. Now $4\cos^2\measuredangle = 4$ iff $\theta_i = \pm\theta_j$. From this i), ii) and iii) easily follow. The proof of iv) is omitted. q.e.d.

Definition. Let G be a connected compact Lie group and fix a maximal torus. choose a fundamental Weyl chamber and write $\Delta = \Delta_+ \cup \Delta_-$. Then a positive root α is called simple if there do not exist $\beta, \gamma \in \Delta_+$ with $\alpha = \beta + \gamma$.

Proposition. (G, T as in the above. Fix a FWC.)

i) Any positive root is a nonegative integral combination of simple roots.

ii) Simple roots are linearly independent.

Proof. Let $\alpha \in \Delta_+$. If α is not simple then $\alpha = \beta + \gamma$ for some $\beta, \gamma \in \Delta_+$. If either of β, γ is not simple then repeat the process. Since $|\Delta_+| < \infty$ this process must terminate. This proves i). To prove ii) we suppose that α, β are distinct simple roots and $\langle \alpha, \beta \rangle > 0$. Then $A_{\alpha\beta} = 2\langle \alpha, \beta \rangle / \langle \alpha, \alpha \rangle$ is a positive integer. This means $\beta - \alpha$ is in the α-root string through β. In particular $\beta - \alpha$ is a root. So either $\beta - \alpha$ or $\alpha - \beta$ is a positive root. Thus either $\beta = (\beta - \alpha) + \alpha$ or $\alpha = (\alpha - \beta) + \beta$ is not simple contradicting the assumption. So far we have shown that if α, β are distinct simple roots then $\langle \alpha, \beta \rangle \leq 0$. Suppose the set of simple roots is linearly dependent, then we can write $n\alpha = \Sigma \lambda_s \theta_s$, where $\alpha, (\theta_s)$ are simple and distinct, n a positive integer, (λ_s) nonnegative integers not all zero. Then $n\langle \alpha, \alpha \rangle = \Sigma \lambda_s \langle \theta_s, \alpha \rangle$. But $n\langle \alpha, \alpha \rangle > 0$ and $\Sigma \lambda_s \langle \theta_s, \alpha \rangle \leq 0$. The contradiction finishes the proof. q.e.d.

It follows that the FWC associated with Δ_+ is given simply by $\{x \in \quad : \theta_s(x) > 0, (\theta_s) \text{ simple roots } \}$.

Proposition. Let G be a simple connected compact Lie group. Fix a maximal torus T. Then the roots Δ span t^*.

Proof. Let $z < t$ denote the Lie algebra of $Z(G)$. Write $t = z \oplus z^\perp$ with respect to any $\mathrm{Ad}(G)$-invariant inner product in g. Also write $t^* = \mathrm{span}\, \Delta \oplus \{\mathrm{span}\, \Delta\}^\perp$. Here we are using the earlier identification $t^* = t$. Now $\{\mathrm{span}\, \Delta\}^\perp = \cap \ker \theta_i, \theta_i \in \Delta$ and $\cap \ker \theta_i = z$. But in §2 we saw that G as in the above $Z(G)$ is finite, i.e., $z = 0$. q.e.d.

The following corollary is immediate.

Corollary. Let G be a simple connected compact Lie group. Then the rank of G is equal to the number of simple roots.

Examples.

i) $G = U(n)$,
   $\mathrm{FWC} = \{x = (x_i) : x_1 > x_2 > \ldots > x_n\}, \Delta_+ = \{x_i - x_j : i < j\}$.
   The simple roots are $x_1 - x_2, x_2 - x_3, \ldots, x_{n-1} - x_n$.

ii) $G = Sp(n)$ or $SO(2n + 1; \mathbf{R})$.
   $\mathrm{FWC} = \{x = (x_i) : x_1 > x_2 > \ldots > x_n > 0\}$. The simple roots are $x_1 - x_2, x_2 - x_3, \ldots, x_{n-1} - x_n$, and $2x_n$ for $Sp(n)$, x_n for $SO(2n + 1; \mathbf{R})$.

Definition. Let G be a compact connected Lie group and T, a maximal torus. Fix a FWC. The Dynkin diagram of G is the graph whose vertices (v_i) correspond to the simple roots which is so constructed that v_i and v_j are joined by $A_{ij} \cdot A_{ji}$ arcs. (Here v_i and v_j respectively represent the simple roots θ_i and θ_j. Recall that the possibilities for $A_{ij} \cdot A_{ji}$ are 0,1,2,3.) Given a pair of vertices an arrow is added pointing to the vertex representing the shorter of the two roots if their lengths are different.

One can talk about "abstract" Dynkin diagrams by axiomatizing the properties of the Dynkin diagrams of Lie groups, but we will not go into this. A good reference on this material is [Hu]. The abstract Dynkin diagrams can be classified by purely combinatorial means (e.g., [Hu] pp. 57–63). We omit the proof of the following theorem.

Theorem.

i) Isomorphic (as graphs) Dynkin diagrams yield isomorphic Lie algebras.

ii) The Lie algebra represented by a Dynkin diagram is simple if and only if the Dynkin diagram is connected as a graph.

In view of the preceding theorem one classifies compact connected Lie groups up to Lie algebra. For example, one may consider only the simply connected compact simple Lie groups. The following table gives such a classification.

G	$\dim G$	$\operatorname{rank} G$	$|W(G)|$	
$SU(n)$	$(n-1)(n+1)$	$n-1$	$n!$	$(n \geq 2)^*$
$Spin(2n+1)$	$n(2n+1)$	n	$2^n n!$	$(n \geq 2)^*$
$Sp(n)$	$n(2n+1)$	n	$2^n n!$	$(n \geq 3)^*$
$Spin(2n)$	$n(2n-1)$	n	$2^{n-1} n!$	$(n \geq 4)^*$
E_6	78	6	$2^7 3^2$	
E_7	133	7	$2^{10} 3^4 5 \cdot 7$	
E_8	248	8	$2^{14} 3^5 5^2 7$	
F_4	52	4	$2^7 3^2$	
G_2	14	2	12	

$(^*$ is to avoid repetition.$)$

The Dynkin Diagrams.

$SU(n)$

$Spin(2n+1)$

$Sp(n)$

$Spin(2n)$

E_6

E_7

E_8

F_4

G_2

Chapter II

ALMOST COMPLEX HOMOGENEOUS SPACES

This chapter contains an exposition of the theory of almost complex homogeneous spaces as laid out by Wang [Wan] and Borel-Hirzebruch [B-H].

§7 discusses generalities concerning (compact) homogeneous spaces. In particular, the so-called type I inner symmetric spaces are introduced. In §8 the notion of partial G-flag manifold is defined. Partial G-flag manifolds are the central object of our study.

The next three sections deal with invariant almost complex structures on homogeneous spaces and their integrability. These notions are formulated in terms of the Maurer-Cartan forms of Lie groups and several examples are worked out. In §12 an important notion, that of the horizontal distribution of G/T with G simple, is introduced. The horizontal distribution will play an important role in several of the examples in Chapter IV.

§7. Homogeneous Spaces

Let G be a Lie group and also let H be a closed subgroup. (Recall that the quotient topology on G/H is Hausdorff iff H is closed.) The quotient space $M \cong G/H$ with its natural smooth structure is called a homogeneous space.

Notation. $[g]$ denotes the equivalence class in G/H represented by $g \in G$, i.e., it is the coset gH.

Let $h \in H$. Then $[h] = 0$ will be called the origin in M and we can write $H = G_0$, the isotropy subgroup at $0 \in M$. There is the

27

projection $\pi : G \to M$ given by $g \mapsto [g]$ or $g \mapsto g \cdot 0$ once we think of g as a transformation of M. The following diagram commutes:

$$
\begin{array}{ccc}
G & \xrightarrow{\;L_g\;} & G \\
{\scriptstyle \pi}\downarrow & & \downarrow{\scriptstyle \pi} \\
M & \xrightarrow[\;g\;]{} & M
\end{array}
$$

The above projection is a H-principal fibration and given a representation $\rho : H \to GL(V)$ there appears the associated vector bundle $G \times_H V = G \times V / \sim_H$, where $(g, v)h = (gh, \rho(h^{-1})v)$, $h \in H$.

Equip G with a biinvariant metric, *assuming* its existence. This gives the orthogonal decomposition at the identity $g = h \oplus m$, $m = h^{\perp}$ = the orthogonal complement. $\pi_{*e}|_m$ is a vector space isomorphism and enables us to identify m with $T_0 M$.

Lemma. $\mathrm{Ad}(H)m < m$.

Proof. Remember that the biinvariant metric on G defines an $\mathrm{Ad}(G)$-invariant inner product $\langle , \rangle$ in g. Now $X \in m$ iff $\langle X, Y \rangle = 0$ for every $Y \in h$. Consider $\langle \mathrm{Ad}_h X, Y \rangle$, where $Y \in h$, $X \in m$, $h \in H$. Noting that $\mathrm{Ad}_h \in GL(h)$ we see that there exists $Y' \in h$ with $\mathrm{Ad}_h Y' = Y$. We then have $\langle \mathrm{Ad}_h X, \mathrm{Ad}_h Y' \rangle = \langle X, Y' \rangle = 0$ for every $Y' \in h$. This does it since the choice of Y was arbitrary. q.e.d.

Thus it makes sense to write $\mathrm{Ad}(H)|_m < GL(m)$.

Lemma. $\pi_*(\mathrm{Ad}_h X) = h_* \pi_* X$ for $X \in m$, $h \in H$.

Proof. $\pi_*(\mathrm{Ad}_h X) = \pi_* L_{h*} R_{h^{-1}*} X = h_* \pi_* R_{h^{-1}*} X$ using the above commutative diagram. So it suffices to show that $\pi_* X = \pi_* R_{h^{-1}*} X$. But this is trivial since $R_{h^{-1}}$ induces the identity transformation on $M = G/H$. q.e.d.

The representation $H \to GL(T_0 M)$ given by $h \mapsto h_*$ is called the linear isotropy representation. The above lemma states that the linear isotropy representation upon the identification $T_0 M = m$ is the adjoint representation of H restricted to m. The associated vector bundle $G \times_H m$ is just the tangent bundle TM.

A significant class of examples of homogeneous spaces is furnished by symmetric spaces. Though our principal object of study, namely the G-flag manifolds, are not symmetric spaces they nevertheless play an important role.

Definition. Let G be a Lie group and also let H be a closed subgroup.

i) The homogeneous space G/H is called a symmetric space if there exists an involutive automorphism σ of G ($\sigma^2 = $ id) such that $(H_\sigma)_e < H < H_\sigma$, where H_σ is the fixed point set of σ and $(H_\sigma)_e$ is the identity component of H_σ. We further assume that $\mathrm{Ad}(H)$ ($< GL(\mathfrak{g})$) is compact.

ii) A symmetric space $(G/H, \sigma)$ is said to be inner if σ is an inner automorphism of G.

Examples.

i) $G = U(p+q), H = U(p) \times U(q), \sigma = \mathrm{Inn}_g$, where $g = \begin{bmatrix} I_p & 0 \\ 0 & I_q \end{bmatrix}$. Then $G/H \cong CG_{p+q,q}$, the complex Grassmannian of q-planes in $\mathbf{C}^{p+q}$.

ii) $G = U(n), H = O(n; \mathbf{R}), \sigma = $ the conjugation. Then $(G/H, \sigma)$ is not inner.

We will use the following proposition later.

Proposition. Let G be a connected compact Lie group and also let G/H be an inner symmetric space. Then rank $H = $ rank G, hence H contains a maximal torus of G.

Proof. This follows at once from 5.6 p. 424 [He]. q.e.d.

Any symmetric space is a product of "irreducible" symmetric spaces (pp. 377–381 [He]). Using the classification of simple compact Lie groups one can obtain the following classification (up to finite covering maps) of connected compact irreducible symmetric spaces: $SO(p + q; \mathbf{R})/S(O(p) \times O(q))$, $SO(2n; \mathbf{R})/U(n)$, $SU(p + q)/S(U(p) \times U(q))$, $SU(n)/SO(2n; \mathbf{R})$, $Sp(n)/U(n)$, $Sp(p + q)/Sp(p) \times Sp(q)$, $SU(2n)/Sp(n)$, $SO(n; \mathbf{R})$, $SU(n), Sp(n)$ and 17 exceptional spaces. These are so called types I and II irreducible symmetric spaces (cf. pp. 516–518 [He]).

Remarks.

i) A compact Lie group is naturally a symmetric space by $G \cong G \times G/\text{diagonal}$. See pp. 223–224 [He]. Type II spaces are compact Lie groups.

ii) The remaining two types (types III and IV) are noncompact and may be realized as "duals" to compact types. See §5 Chapter 8 [He].

iii) The complete list of irreducible compact hermitian symmetric spaces is: $SO(n+2;\mathbf{R})/SO(2) \times SO(n)$, $SO(2n;\mathbf{R})/U(n)$, $Sp(n)/U(n)$, $SU(p+q)/S(U(p) \times U(q))$, $E_6/Spin(10) \times T^1$ (EIV), $E_7/E_6 \times T^1$ (EVII).

§8. Partial G-Flag Manifolds and Invariant Metrics

Definition. Let G be a connected compact semisimple Lie group and also let T be a maximal torus in G. The G-flag manifold is defined to be the homogeneous space G/T. We also define a partial G-flag manifold to be any homogeneous space of the form G/H, where H is a connected closed subgroup of maximal rank.

In the above definition we can assume that $H > T$, replacing T by one of its conjugates if necessary. Then there is the projection $G/T \to G/H$ given by $gT \mapsto gH$.

Definition. A metric $ds^2 = \langle\,,\,\rangle$ on $M = G/H$ is said to be invariant if $\langle X, Y \rangle = \langle g_* X, g_* Y \rangle$ for every $X, Y \in T_p M$, $g \in G$.

Recall that a homogeneous space G/H is said to be reductive if there exists a subspace $m < g$ complementary to h with $\text{Ad}(H)m < m$. Note also that a partial G-flag manifold is reductive.

Proposition. Let $M = G/H$ be any reductive homogeneous space and also let $g = h \oplus m$ be a vector space direct sum decomposition with $\text{Ad}(H)m < m$. Then the identification $m = T_0 M$ (via the vector space isomorphism π_{*e}) gives rise to a bijective correspondence between $\text{Ad}(H)$-invariant inner products in m and invariant metrics on M.

Proof. Let $\langle , \rangle$ be an invariant metric on G/H. Then $\langle X, Y \rangle_0 = \langle h_* X, h_* Y \rangle_0$, $h \in H, X, Y \in T_0 M$. But upon the identification $m = T_0 M$, h_* becomes Ad_h. Conversely let $(,)$ denote an $\text{Ad}(H)$-invariant inner product in m. Identify m with $T_0 M$ and put $\langle , \rangle_0 = (,)$. Let $p = g(0) = g'(0) \in M$, $g, g' \in G$. Then we must have $g'^{-1} g \in H$. Let $v, w \in T_p M$. Then there exist uniquely determined $X, Y, X', Y' \in m$ such that $g_* X = g'_* X' = v$, $g_* Y = g'_* Y' = w$ since g_* and g'_* are isomorphisms. Define $\langle v, w \rangle_p = \langle X, Y \rangle_0$. We need to show that $\langle X, Y \rangle_0 = \langle X', Y' \rangle_0$. Now $g_* X = g'_* X'$ imples that $g'^{-1}_* g_* X = h_* X = X'$ for some $h \in H$. Likewise $Y' = h_* Y$. So $\langle X', Y' \rangle_0 = \langle h_* X, h_* Y \rangle_0 = (\text{Ad}_h X, \text{Ad}_h Y) = (X, Y) = \langle X, Y \rangle_0$. q.e.d.

In §2 we saw that a connected semisimple Lie group is compact iff its Cartan-Killing form κ is negative definite. Note also that κ is easily $\text{Ad}(G)$-invariant. Let G/H be a partial G-flag manifold. Then $-\kappa$ (or any negative multiple of κ) provides an $\text{Ad}(G)$-invariant inner product in g. With respect to this inner product we put $m = h^\perp$. We saw earlier (§7) that $\text{Ad}(H)m < m$. It follows that $-\kappa|_m$ is an $\text{Ad}(H)$-invariant inner product. Now the preceding lemma says that $-\kappa|_m = -\kappa|_{T_0 M}$ is uniquely extended to produce an invariant metric on M. From now on when we speak of a Riemannian structure on a partial G-flag manifold we always use the above invariant metric (or a scalar multiple of it.)

We know that for a connected compact simple Lie group any two biinvariant metrics are constant multiples of each other. This fact enables us to obtain the totality of invariant metrics on G/T with G simple, as we shall show in the following: Recall the root space decomposition $g = t \oplus \Sigma V_i$. Note that here the root spaces (V_i) are independent of the choice of $\text{Ad}(G)$-invariant inner product in   since any other $\text{Ad}(G)$-invariant inner product in g is just a multiple of the given one. Thus we can write $m = \oplus \Sigma V_i = t^\perp$ without really choosing an inner product. Note that we also have $\text{Ad}(T)m < m$. Now an invariant metric on $M = G/T$ corresponds to an $\text{Ad}(T)$-invariant inner product in m. $\text{Ad}(T)$ restricted to V_i is irreducible and it is easily seen that V_i possesses only a one-dimensional family of $\text{Ad}(T)$-invariant inner products. This gives

Theorem. (Maintain the above notation.)

Let $M = G/T$ be a G-flag manifold with G simple. Then the totality of $\mathrm{Ad}(T)$-invariant inner products in m (hence the totality of invariant metrics on M) is given by $\{\oplus\Sigma\, c_i \cdot \kappa|_{V_i} : (c_i)$ all negative $\}$, where κ is the Cartan-Killing form of G.

§9. Invariant Complex Structures

Definition. An invariant almost complex structure on a homogeneous space $M = G/H$ is a smooth field of complex structures, denoted by J, on TM such that $J \circ g_* = g_* \circ J$ for every $g \in G$.

A homogeneous space equipped with an invariant almost complex structure is called an almost complex homogeneous space.

Proposition. Let $M = G/H$ be a reductive homogeneous space and choose a vector space direct sum decomposition $g = h \oplus m$ with $\mathrm{Ad}(H)m < m$. Then the identification $T_0 M = m$ via π_{*e} induces a bijective correspondence between invariant almost complex structures on M and $\mathrm{Ad}(H)$-invariant complex structures on m.

Proof. Let J be an invariant almost complex structure on M. Then J at the origin, denoted by J_0, is a complex structure on $T_0 M$, hence can be viewed as a complex structure on m using the identification via π_{*e}. Now J is invariant, so J_0 commutes with the linear isotropy representation in particular. But under the identification $T_0 M = m$ the linear isotropy representation becomes the adjoint representation of H on m as we saw in §7. Conversely let J' be an $\mathrm{Ad}(H)$-invariant complex structure on m. We can view J' as a complex structure on $T_0 M$. Put $J_0 = J'$ and for $x \in M$ define J_x by the formula $J_x v = g_*^{-1} J_0 g_* v$, where $g(x) = 0$. J_x is well-defined: for if $g(x) = g'(x) = 0$ then $gg'^{-1} \in H$. Now by the $\mathrm{Ad}(H)$-invariance of J' we get $g_* g_*'^{-1} J_0(g_*' v) = J_0 h_*(g_*' v) = J_0 g_* v$. Thus $g_* g_*'^{-1} J_0 g_*' v = J_0 g_* v$ and this gives $g_*'^{-1} J_0 g_*' v = g_*^{-1} J_0 g_*' v$. Invariance of J follows from its construction. q.e.d.

Given a complex manifold the multiplication by $\sqrt{-1}$ defines an almost complex structure on it. So a complex manifold is naturally an almost complex manifold. However, it is not true that any almost complex manifold can be made into a complex manifold. A deep theorem of Newlander-Nirenberg states that an almost complex manifold can be made into a complex manifold if it possesses a so-called integrable almost complex structure. It is also trivially true that the natural almost complex structure on a complex manifold is integrable. (Actually the above theorem reduces to an elementary application of the Fröbenius theorem on exterior differential systems if one restricts to the class of real analytic almost complex structures and we primarily work in the real analytic category.) For a detailed account of the notion of integrability we refer the reader to [Ch3] pp. 12–17. In §10 we will give a rigorous definition of the integrability in the context of partial G-flag manifolds.

We now mention two related theorems of Wang which serve as a motivation for studying partial G-flag manifolds.

Theorem (Wang). Let G/H be a partial G-flag manifold. Then there exists an integrable invariant almost complex structure on G/H if and only if H is the centralizer of a torus in G.

For a proof see [B-H] pp. 501–502. Also see [Wan].

Theorem (Wang). Any simply connected compact complex homogeneous space is the base space of a toral fibre bundle (with possibly trivial fibre) $P \to M$, where P is the product of partial G-flag manifolds with G's all simple and H's the centralizers of tori in G's.

See [Wan] pp. 29–31 for a proof.

Let G/T be a G-flag manifold and G/H, a partial G-flag manifold with $H > T$. On G we use the biinvariant metric coming from the negative of κ, the Cartan-Killing form of G. At the identity we then have the orthogonal decomposition $g = t \oplus t^\perp$ and $t^\perp = \oplus \Sigma V_i$, where V_i's are the root spaces discussed in §4. There is also the orthogonal decomposition $g = h \oplus m$, $m = h^\perp$. Since $H > T$ we must have $m < t^\perp$. We have

Proposition. $h = t \oplus \Sigma V_j, j \in J < I$ and $m = \oplus \Sigma V_a, a \in I \backslash J$.

Proof. Clearly one equality implies the other and we prove the latter. Let $m = \oplus \Sigma n_\alpha$ be the irreducible decomposition of m into $\mathrm{Ad}(H)$-invariant subspaces. This makes sense since $\mathrm{Ad}(H)m = m$. Since

$H > T$ it follows that (n_α) are also $\mathrm{Ad}(T)$-invariant subspaces. But (V_i) are all irreducible $\mathrm{Ad}(T)$-invariant subspaces and $\oplus \Sigma V_i > m$. q.e.d.

Observe that (V_j) are the root spaces of H relative to T and the inclusion $J < I$ signifies the inclusion of the roots $\Delta^H < \Delta^G$ using the obvious notation. We write $\Delta^G = \{\pm\theta_i : i \in I\}$, $\Delta^H = \{\pm\theta_j : j \in J\}$.

Definition. $\Delta^M = \Delta^G \backslash \Delta^H = \{\pm\Theta_a : a \in I\backslash J\}$ is called the set of complementary roots of G to H relative to T.

Theorem. (Maintain the above notation.)

There exist exactly $2^m (m = |I|)$ many invariant almost complex structures on G/T.

Proof. We recall some notation from §4. On each root space V_i we had orthonormal basis $\{E_{2i-1}, E_{2i}\}$ and relative to this basis $\rho_i = \mathrm{Ad}(T)|_{V_i} < SO(2; \mathbf{R})$. Observe that since ρ_i is nontrivial there exist precisely two complex structures on V_i commuting with ρ_i represented by the matrices $\pm \begin{pmatrix} 0 & -1 \\ 1 & 0 \end{pmatrix}$ written relative to the orthonormal basis $\{E_{2i-1}, E_{2i}\}$. We denote these complex structures by $\pm J_i$. Now an invariant almost complex structure on G/T is identified, by restriction, with an $\mathrm{Ad}(T)$-invariant complex structure on $\oplus \Sigma V_i$. But any $\mathrm{Ad}(T)$-invariant complex structure on $\oplus \Sigma V_i$ must leave the invariant subspaces (V_i) fixed and on each V_i there are exactly two invariant complex structures $\pm J_i$. q.e.d.

Let J be an invariant almost complex structure on the G-flag manifold G/T. Then using the above proof it makes sense to write $J = \oplus \Sigma \varepsilon_i J_i, i \in I, \varepsilon_i = \pm 1$.

For a partial G-flag manifold in general there may not exist any invariant almost complex structure, as we shall see below. Let $M = G/H$ be a partial G-flag manifold and recall the usual orthogonal decomposition $g = h \oplus m$, say relative to $-\kappa$. Upon identifying $T_0 M$ with m the linear isotropy representation of M becomes $\mathrm{Ad}(H)|m < GL(m)$. Noting that $-\kappa|_m$ is $\mathrm{Ad}(H)$-invariant we obtain an orthogonal matrix representation $\mathrm{Ad}(H)|_m < SO(2n; \mathbf{R})(2n = \dim M)$ by choosing an orthonormal basis of m. Let $i : H \to SO(2n; \mathbf{R})$ denote this orthogonalized linear isotropy representation.

Proposition. (Maintain the preceding notation.)

There exists an invariant almost complex structure on G/H if and only if $i(H)$, possibly rechoosing an orthonormal basis for m, lies in the standard inclusion of $U(n)$ in $SO(2n; \mathbf{R})$ $\left(A + iB \mapsto \begin{pmatrix} A & -B \\ B & A \end{pmatrix} \right)$.

Proof. Suppose there exists an invariant almost complex structure on G/H. This means there exists an $\mathrm{Ad}(H)$-invariant complex structure J on m. Since $T < H, J$ is also $\mathrm{Ad}(T)$-invariant and consulting the proof of the preceding theorem we see that J is of the form $\oplus \Sigma \, \varepsilon_a J_a, \varepsilon_a = \pm 1$ written relative to an orthonormal basis. (Of course not every choice of (ε_a) represents an $\mathrm{Ad}(H)$-invariant complex structure unlike the case of G/T.) Recording the basis vectors put $J = \begin{bmatrix} 0 & -I_n \\ I_n & 0 \end{bmatrix}$. Wlog we may assume that with respect to the same basis $i : H \to SO(2n; \mathbf{R})$. Now $J \circ i_h = i_h \circ J$ iff i_h is in the standard inclusion of $GL(n; \mathbf{C}) < GL(2n; \mathbf{R})$. But this means $i(H)$ is in the standard inclusion of $U(n) < SO(2n; \mathbf{R})$. Conversely suppose that $i(H)$ lies in the inclusion of $U(n) < SO(2n; \mathbf{R})$. Then the matrix $\begin{bmatrix} 0 & -I_n \\ I_n & 0 \end{bmatrix}$ commutes with elements of $i(H)$, hence represents an $\mathrm{Ad}(H)$-invariant complex structure on m. q.e.d.

Corollary. There does not exist an invariant almost complex structure on $HP^n = Sp(n+1)/Sp(1) \times Sp(n)$.

Proof. We have $i : Sp(1) \times Sp(n) \to SO(4n; \mathbf{R})$, $4n = \dim HP^n$. Take $(a, A) \in Sp(1) \times Sp(n), q = {}^t(q^1, \ldots, q^n) \in \mathbf{H}^n$. Computing the adjoint representation we get $i(a, A) : q \mapsto A^{-1}qa$. Observe that $i(1 \times Sp(n)) = Sp(n)$. It follows that $\dim i(Sp(1) \times Sp(n)) \geq \dim Sp(n) = 2n^2 + n$. Now suppose there exists an invariant almost complex structure on HP^n. Then by the above proposition we must have $i(Sp(1) \times Sp(n)) < U(2n)$. But this would imply that $\dim U(2n) = 2n^2 \geq \dim i(Sp(1) \times Sp(n)) \geq 2n^2 + n$ and the contradiction finishes the proof. q.e.d.

Given a partial G-flag manifold G/H we let $m = \oplus \Sigma \, n_\alpha, \alpha = 1$ to n', be the irreducible decomposition of m into its $\mathrm{Ad}(H)$-invariant subspaces. Recalling the containment $H > T$ we see that each n_α is a direct sum of $Ad(T)$-invariant subspaces V_i's. We then have

Proposition. Suppose G/H has an invariant almost complex structure. Then there exist precisely $2^{n'}$ many such structures on G/H.

For a proof see p. 501 of [B-H]. The upshot here is that on each n_α there are exactly two $Ad(H)$-invariant complex structures differing by sign only.

Note that we can still use the notation $J = \oplus \Sigma \varepsilon_a J_a (a \in \Gamma \backslash J, \varepsilon_a = \pm 1)$ for a given $Ad(H)$-invariant complex structure on m. Of course, in general $|\Gamma \backslash J| >> n'$ and an arbitrarily chosen such expression does not represent an $Ad(H)$-invariant complex structure on m.

Definition. Let $J = \oplus \Sigma \varepsilon_a J_a$ represent an invariant almost complex structure on a partial G-flag manifold $M = G/H$. Then the roots $\{\varepsilon_a \theta_a : a \in \Gamma \backslash J\} < \Delta^M$ are called the roots of the almost complex structure J.

§10. The Maurer-Cartan Form

Definition. Let G be any Lie group. The Maurer-Cartan form of G is a g-valued 1-form on G given by $\Omega : TG \to g\, T_e(G)$, $\Omega_g(X) = L_{g^{-1}*} X, X \in T_g G$.

Observations.

i) Ω is left invariant: $L_g^{*'} \Omega_g(X) = \Omega_g(L_{g'*} X) = L_{(g'g)^{-1}*} L_{g'*} X = L_{g^{-1}*} X = \Omega_g(X)$.

ii) For $G = GL(n; \mathbf{R})$ we have the following: Let $X = (X_j^i) : G \to \mathbf{R}^{n^2}$ denote the usual matrix coordinates. Writing $dX = (dX_j^i)$ we get $\Omega = X^{-1} dX$. The exterior differentiation of the equation $dX = X\Omega$ leads to $d\Omega = -\Omega \wedge \Omega$ which (written out in components) is the so-called structure equations of G.

iii) For a closed subgroup $G < GL(n; \mathbf{R})$, Ω and the structure equations are obtained simply by restrictions.

Let G be a connected compact Lie group and T, a maximal torus in G. Then there is the root space decomposition $g = t \oplus \Sigma V_i$ and it gives rise to the decomposition $\Omega = \Omega_t + \Sigma \Omega_{V_i}$.

Examples.

i) $G = SO(2n; \mathbf{R}), T = SO(2)^n$.
$t^\perp = \oplus \Sigma V_{ij} \oplus \Sigma V'_{ij} (1 \le i < j \le n)$ where $V_{ij} = \mathbf{R}$-span $\{E_{ij}, F_{ij}\}$ and $V'_{ij} = \mathbf{R}$-span $\{E'_{ij}, F'_{ij}\}$. (See Example iv) §4 for the notation.) $\Omega = X^{-1}dX$ where $X = (X^i_j) : SO(2n; \mathbf{R}) \to \mathbf{R}^{4n^2}$ the usual coordinates. We obtain

$$\Omega_{V_{ij}} = \frac{1}{2}[(\Omega^{2i-1}_{2j-1} + \Omega^{2i}_{2j}) \otimes E_{ij} + (\Omega^{2i}_{2j-1} - \Omega^{2i-1}_{2j}) \otimes F_{ij}],$$

$$\Omega_{V'_{ij}} = \frac{1}{2}[(\Omega^{2i-1}_{2j-1} - \Omega^{2i}_{2j}) \otimes E'_{ij} + (\Omega^{2i}_{2j-1} + \Omega^{2i-1}_{2j}) \otimes F'_{ij}] \ (\text{no sum}).$$

ii) $G = U(n), T = U(1)^n$.
$t^\perp = \oplus \Sigma V_{ij} (1 \le i < j \le n)$, where $V_{ij} = \mathbf{R}$-span $\{E_{ij}, F_{ij}\}$. (See Example i) §4 for the notation.) Let $Z = (Z^i_j) = (X^i_j + \sqrt{-1}Y^i_j)$: $U(n) \to \mathbf{C}^{n^2} = \mathbf{R}^{2n^2}$ the usual coordinates. Write $\Omega = Z^{-1}dZ$ and put $\Omega^i_j = \Omega'^i_j + \sqrt{-1}\Omega''^i_j$, where Ω' and Ω'' are real valued. Then $\Omega_{V_{ij}} = \Omega'^i_j \otimes E_{ij} + \Omega''^i_j \otimes F_{ij}$ (no sum).

The Maurer-Cartan form of G is useful in expressing invariant metrics on partial G-flag manifolds. We first look at the following standard result on Cartan-Killing forms of classical groups.

Proposition. Let $G = O(n; \mathbf{R})(n > 2), U(n)(n > 1)$ or $Sp(n)$ where we think of $Sp(n)$ as a closed subgroup of $GL(n; \mathbf{C})$ via the usual injection. Then $\kappa(X, Y) = c(\operatorname{tr} XY)$ with $c = n - 2$ for $O(n; \mathbf{R})$, $2n$ for $U(n), 2n + 2$ for $Sp(n)$.

Proof. Keeping in mind the usual tangent space identification for matrix groups we get $\operatorname{Ad}_g X = gXg^{-1}$ for $g \in G, X \in \mathfrak{g}$. The rest is purely computational. q.e.d.

Corollary. G as in the above proposition $\kappa = c \cdot \operatorname{tr}(\Omega^2)$.

Examples.

i) $G = U(n)$.
$\Omega = Z^{-1}dZ$, the Maurer-Cartan form of G. The negative of the Cartan-Killing form gives rise to the biinvariant metric on G given by $(-2n)\operatorname{tr}(\Omega^2)$. Let $H = U(1) \times U(n-1) > T = U(1)^n$ and also let $M = G/H$. We have $\mathfrak{g} = \mathfrak{h} \oplus \mathfrak{m}$ and $\Omega = \Omega_{\mathfrak{h}} \oplus \Omega_m, \Omega_m =$

$\Sigma(\Omega'^1_i \otimes E_{1i} + \Omega''^1_i \otimes F_{1i}), 2 \leq i \leq n$. Of course the pullback here is independent of the choice of a local section of $G \to M$ by its very construction. The above obtained metric on $M = CP^{n-1}$ is called the Fubini-Study metric.

ii) $G = SO(2n; \mathbf{R})$.

$\Omega = X^{-1}dX$ as before. The negative of the Cartan-Killing form, upon translation, yields the biinvariant metric $(2 - 2n)\mathrm{tr}(\Omega^2)$ on G. Let $H = SO(2) \times SO(2n - 2) < T = SO(2)^n$ and also let $M = G/H$. We then obtain

$$\Omega_m = \begin{bmatrix} 0 & & * \\ \Omega_1^3, \Omega_2^3 & & \\ \vdots & & 0 \\ \Omega_1^{2n}, \Omega_2^{2n} & & \end{bmatrix}, * = -\text{transpose}.$$

The resulting invariant metric on $M = Q_{n-1}$ is given by the pullback of $2(2n - 2)\Sigma(\Omega_1^i)^2 + (\Omega_2^i)^2, 3 \leq i \leq 2n$.

Let $M = G/H$ be a partial G-flag manifold endowed with an invariant almost complex structure J. Recall the orthogonal decomposition $g = h \oplus m$ and also that J is identified with an $\mathrm{Ad}(H)$-invariant complex structure on m. Complexify $m, m_C = m \otimes C$. J gives rise to the eigenspace decomposition $m_C = m^+ \oplus m^-$ (or sometimes we write $m^{(1,0)} \oplus m^{(0,1)}$) of m_C into $\pm\sqrt{-1}$-eigenspaces of $J_C : m_C \to m_C$ (J_C = the linear extension of J).

Let $m = \oplus\Sigma V_a(a \in I\backslash J$, see §9 for the notation) be the root space decomposition of m. Choose an orthonormal basis $\{E_{2a-1}, E_{2a}\}$ for each V_a (and hence an orthonormal basis for m) and write $J|_{V_a} = \varepsilon_a \begin{pmatrix} 0 & -1 \\ 1 & 0 \end{pmatrix}, \varepsilon_a = +1$ or -1. A little computation gives

Lemma. $E_{2a-1} - i\varepsilon_a E_{2a} \in m^{(1,0)}, E_{2a-1} + i\varepsilon_a E_{2a} \in m^{(0,1)}$.

Write $\Omega_m = \Omega^{2a-1} \otimes E_{2a-1} + \Omega^{2a} \otimes E_{2a}$ for some ordinary 1-forms $\Omega^{2a-1}, \Omega^{2a}(a \in I\backslash J)$.

Notation. $\Omega_m^{(1,0)} = \Omega^{2a-1} \otimes E_{2a-1} + i\varepsilon_a\Omega^{2a} \otimes E_{2a},$
$\Omega_m^{(0,1)} = \Omega^{2a-1} \otimes E_{2a-1} - i\varepsilon_a\Omega^{2a} \otimes E_{2a},$
$\Theta^a = \Omega^{2a-1} + i\varepsilon_a\Omega^{2a}, a \in I\backslash J.$

Proposition. Let $M = G/H$ be a partial G-flag manifold equipped with an invariant almost complex structure J. Then the type $(1, 0)$ forms (respectively, type $(0, 1)$ forms) on M are given by C-linear combinations of the pullbacks (by local sections of $G \to M$) of (Θ^a) (respectively, $(\bar{\Theta}^a)$).

Proof. Note that (Θ^a) annihilate $m^{(0,1)}$ at the identity. The invariance of J and the left invariance of Ω do the rest. q.e.d.

Examples.

i) $G = U(n), H > T = U(1)^n$.

We have $g = t \oplus \Sigma V_{ij} (1 \leq i << n)$, where $V_{ij} = \mathbf{R}$-span $\{E_{ij}, F_{ij}\}$. (See §4 for the notation.) $\Omega_{V_{ij}} = \Omega_j'^i \otimes E_{ij} + \Omega''^i_j \otimes F_{ij}$ (no sum), where $\Omega_j'^i = \frac{1}{2}(\Omega_j^i - \Omega_i^j) = \operatorname{Re} \Omega_j^i, \Omega''^i_j = -\frac{i}{2}(\Omega_j^i + \Omega_i^j) = \operatorname{Im} \Omega_j^i$ and $(\Omega_j^i) = Z^{-1}dZ$ as before. We take the invariant complex structure J with $\varepsilon_{ij} = +1$ for all complementary indices (ij). (Other cases are completely analogous as it will become obvious.) C-span $\{E_{ij}, F_{ij}\} = V_{ij}^+ \oplus V_{ij}^-$. Now $v = \binom{x}{y} \in V_{ij}^+$ iff $\begin{pmatrix} 0 & -1 \\ 1 & 0 \end{pmatrix} \binom{x}{y} = i\binom{x}{y}$ iff $v = x(E_{ij} - \sqrt{-1}F_{ij}) = \binom{x}{-ix}, x \neq 0$. Likewise $v = \binom{x}{y} \in V_{ij}^-$ iff $v = x(E_{ij} + \sqrt{-1}F_{ij})$. Consider C-span $\{\Omega_j'^i, \Omega''^i_j\}$. Let $\Phi = a\Omega_j'^i + b\Omega''^i_j, a, b$ complex. Set $\Phi(v) = 0$ for $v \in V_{ij}^-$. We then must have $a = b/i$ and hence the pullback of $\Omega_j'^i + \sqrt{-1}\Omega''^i_j = \Omega_j^i$ is a type $(1, 0)$ form on G/H. If, for example, $H = U(1) \times U(n - 1)$ then the type $(1, 0)$ forms on M are C-linear combinations of the pullbacks of Ω_i^1 if $\varepsilon_{1i} = 1$ and $\bar{\Omega}_i^1$ if $\varepsilon_{1i} = -1, 2 \leq i \leq n$.

ii) $G = SO(2n; \mathbf{R}), H > T = SO(2n)^n$.

Again we maintain the notation introduced in §4. Let $I \backslash J$ denote the appropriate complementary index set for G/H. Equip G/H with the invariant almost complex structure arising from the choice $\{\varepsilon_{ij}(= \pm 1) : (ij) \in I \backslash J\}$. Then the type $(1, 0)$ forms on G/H are C-linear combinations of the pullbacks of

$$\Theta^{ij} = \frac{1}{2}[(\Omega_{2j-1}^{2i-1} + \Omega_{2j}^{2i}) + \sqrt{-1}\varepsilon_{ij}(\Omega_{2j-1}^{2i} - \Omega_{2j}^{2i-1})],$$

$$\Theta'^{ij} = \frac{1}{2}[(\Omega_{2j-1}^{2i-1} - \Omega_{2j}^{2i}) + \sqrt{-1}\varepsilon_{ij}(\Omega_{2j-1}^{2i} + \Omega_{2j}^{2i-1})].$$

Remark. Let $M = G/H$ be a partial G-flag manifold. At the identity there are the orthogonal decompositions $g = t \oplus \Sigma V_i (i \in I) = h \oplus \Sigma V_a (a \in I\backslash J)$. Let $\{E_{2a-1}, E_{2a} : a \in I\backslash J\}$ be an orthonormal basis for $m = \oplus \Sigma V_a$ with $\mathbb{R}$-span $\{E_{2a-1}, E_{2a}\} = V_a$ for every a. Then any other such basis $\{\tilde{E}_{2a-1}, \tilde{E}_{2a} : a \in I\backslash J\}$ of m is given by

$$(\tilde{E}_{2a-1}, \tilde{E}_{2a}) = (E_{2a-1}, E_{2a}) \begin{pmatrix} \cos\alpha_a & -\sin\alpha_a \\ \sin\alpha_a & \cos\alpha_a \end{pmatrix},$$

or using the complex notation

$$\tilde{E}_{2a-1} + i\tilde{E}_{2a} = e^{-i\alpha_a}(E_{2a-1} + iE_{2a}).$$

That is to say, the basis $\{E_{2a-1}, E_{2a} : a \in I\backslash J\}$ is determined up to the structure group $SO(2;\mathbb{R}) \times \ldots \times SO(2;\mathbb{R}) = U(1) \times \ldots \times U(1)$. It follows that $\Omega^{2a-1} + i\Omega^{2a} = e^{i\alpha_a}(\tilde{\Omega}^{2a-1} + i\tilde{\Omega}^{2a})$ and $\Omega^{2a-1} - i\Omega^{2a} = e^{-i\alpha_a}(\tilde{\Omega}^{2a-1} - i\tilde{\Omega}^{2a})$. At any rate we have $\Theta^a = e^{i\epsilon_a \alpha_a}\tilde{\Theta}^a$.

§11. Integrability Condition

Let $M = G/H$ be a partial G-flag manifold equipped with an invariant almost complex structure J. Recall that M is a Riemannian reductive homogeneous space and at the identity of G there is the orthogonal decomposition $g = h \oplus m, m = T_0 M$. As we saw earlier m is further decomposed into two dimensional root spaces $m = \oplus \Sigma V_a, a \in I\backslash J$. (Remember that $I\backslash J$ is the complementary index set. See §9.) J determines the forms $\Theta = (\Theta^a = \Omega^{2a-1} + \sqrt{-1}\epsilon_a \Omega^{2a})$ up to the structure group $U(1) \times \ldots \times U(1)$ on G. And we saw that C-span of the pullbacks of these forms to M gives the type $(1, 0)$ forms on M.

Definition. Almost complex structure J on $M = G/H$ is said to be integrable if $d\Theta^a \equiv O(\text{mod } \Theta)$ on M.

On M (i.e., pulling back forms to M) $(\Theta^a, \bar{\Theta}^a)$ form a pointwise C-basis for the complexified cotangent bundle T^*M. Therefore we have

$$d\Theta^a = \frac{1}{2}A^a_{bc}\Theta^b \wedge \Theta^c + B^a_{bc}\Theta^b \wedge \bar{\Theta}^c + \frac{1}{2}C^a_{bc}\bar{\Theta}^b \wedge \bar{\Theta}^c$$

for some complex valued functions A, B, C with A_{bc}, C_{bc} skew symmetric. Then J is integrable iff $(C^a_{bc}) = 0$.

The above formulation of integrability has the advantage of computability by virtue of the structure equations $d\Omega = -\Omega \wedge \Omega$. We shall exhibit this in the following example.

Example. $G = U(n), H = U(1) \times U(n-1)$.

Let J correspond to $\{\varepsilon_{1i} : 2 \leq i \leq n\}$. Then $\Theta^i = \Omega_i^1$ if $\varepsilon_{1i} = 1$ and $\bar{\Omega}_i^1$ if $\varepsilon_{1i} = -1$. Now

$$d\Omega_i^1 = -\Omega_1^1 \wedge \Omega_i^1 - \Omega_j^1 \wedge \Omega_i^j \quad (j > 1),$$
$$d\bar{\Omega}_i^1 = -\bar{\Omega}_1^1 \wedge \bar{\Omega}_i^1 - \bar{\Omega}_j^1 \wedge \bar{\Omega}_i^j \quad (j > 1).$$

So clearly if $\varepsilon_{1i} = +1$ for every i or if $\varepsilon_{1i} = -1$ for every i then the integrability condition is met. On the other hand let us suppose that we have mixed indices, say $\varepsilon_{1a} = -1$ for some a and $\varepsilon_{1b} = +1$ for some b. Then

$$d\Theta^b = d\Omega_b^1 = -\Omega_1^1 \wedge \Omega_b^1 - \Omega_j^1 \wedge \Omega_b^j (j > 1) = -\Omega_1^1 \wedge \Theta^b - \overset{\downarrow}{\Omega}_j^1 \wedge \Omega_b^j,$$

and the arrowed term would contain some $\bar{\Theta}^a$ and corresponding J can not be integrable.

Definition. A set $S < \Delta$ of roots of G is said to be closed if for every $\alpha, \beta \in S$ with $\alpha + \beta \in \Delta$ we have $\alpha + \beta \in S$.

For example, a system of positive roots is always closed.

The following proposition gives a "dual" description of the integrability. Its proof is fairly starightforward and can be found in [B-H] p. 499.

Proposition. An invariant almost complex structure on $M = G/H$ is integrable if and only if the set of its roots (as defined in §9) is closed.

Corollary. Let $M = G/T$ be a G-flag manifold and also let $W(G)$ denote the Weyl group of G. Then there are exactly $|W(G)|$ many integrable invariant almost complex structures on M.

Proof. Let J denote an invariant almost complex structure on M. (We have already shown the existence of J.) Then the roots of J is closed iff they actually form a system of positive roots of G. Now a system of positive roots corresponds to a Weyl chamber and in §5 we saw that $W(G)$ acts simply transitively on the set of Weyl chambers. q.e.d.

It is not difficult to see that the above complex structures are all related by diffeomorphisms of G/T induced from inner automorphisms of G coming from the Weyl group action. However, we shall keep these structures separate as it is essential to do so when one considers certain composition of maps $S \to G/T \to G/H$.

Examples.

i) The following spaces are almost complex but never complex ([B-H] pp. 500-501): $G_2/SU(3) \cong S^6, F_4/SU(3) \times SU(3), E_6/SU(3) \times SU(3) \times SU(3), E_7/SU(3) \times SU(6), E_8/SU(3) \times E_6, E_8/SU(9), E_8/SU(5) \times SU(5)$.

ii) The complex flag manifold $SU(n)/S(U(1)^n)$ has $n! = |W(SU(n))|$ many integrable invariant almost complex structures. On the other hand it possesses $2^{\frac{1}{2}(n^2-n)}$ many invariant almost complex structures. Let $G = SU(3), T = S(U(1)^3)$. Then there are $2^3 = 8$ invariant almost complex structures and exactly $3! = 6$ of them are integrable. Maintaining the usual notation we give the complete list of such structures. (Verifications are easy and left to the reader.)

	J_1	J_2	J_3	J_4	J_5	J_6	J_7	J_8
Θ^{12}	Ω_2^1	$\bar\Omega_2^1$	Ω_2^1	Ω_2^1	Ω_2^1	$\bar\Omega_2^1$	$\bar\Omega_2^1$	$\bar\Omega_2^1$
Θ^{13}	Ω_3^1	$\bar\Omega_3^1$	Ω_3^1	$\bar\Omega_3^1$	$\bar\Omega_3^1$	Ω_3^1	Ω_3^1	$\bar\Omega_3^1$
Θ^{23}	Ω_3^2	Ω_3^2	Ω_3^2	Ω_3^2	Ω_3^2	Ω_3^2	Ω_3^2	Ω_3^2

All except J_4 and J_7 are integrable.

iii) Hermitian symmetric spaces.

Let $M = SU(p + q)/S(U(p) \times U(q)) = G/H$. Choose a FWC such that $\Delta_+^G = \{-x_i + x_j : 1 \leq i < j \leq p+q\}$. Then $\Delta_+^H = \{-x_i + x_j : 1 \leq i < j \leq p\} \cup \{-x_i + x_j : p+1 \leq i < j \leq p+q\}$. So the complementary positive roots are $\Delta_+^M = \{-x_i + x_j : 1 \leq i \leq$

$p, p+1 \leq j \leq p+q$. Then Δ_+^M is closed and represents the roots of an integrable invariant almost complex structure. (Here we could have just as easily used the Maurer-Cartan form formulation. The resulting notation would have been somewhat tedious, however.) Observe that the only other choice of the roots of an integrable structure on M is given by $\tilde{\Delta}_+^M = -\Delta_+^M = \{x_i - x_j : 1 \leq i \leq p, p+1 \leq j \leq p+q\}$. The two integrable structures obtained are conjugate to each other. Completely analogous analyses apply to all other hermitian symmetric spaces.

iv) $G = U(4), H = U(2) \times U(1) \times U(1), T = U(1)^4$.

Then $\Delta^G = \{\pm(x_i - x_j) : 1 \leq i < j \leq 4\}$ and $\Delta^H = \{\pm(x_1 - x_2)\}$. Put $\Delta_+^M = \{x_4 - x_1, x_4 - x_2, x_4 - x_3, x_1 - x_3, x_2 - x_3\}$ and $\tilde{\Delta}_+^M = \{x_1 - x_3, x_1 - x_4, x_2 - x_3, x_2 - x_4, x_3 - x_4\}$. Δ_+^M gives $\Theta^{14} = \bar{\Omega}_4^1, \Theta^{24} = \bar{\Omega}_4^2, \Theta^{34} = \bar{\Omega}_4^3, \Theta^{13} = \Omega_3^1, \Theta^{23} = \Omega_3^2$ and $\tilde{\Delta}_+^M$ gives $\Theta^{ij} = \Omega_j^i$ for all i, j. Now Δ_+^M and $\tilde{\Delta}_+^M$ are both closed and define integrable structures on M. However, one can show that the resulting complex structures are not equivalent under Diffeo(M). See [B-H] p. 505.

§12. The Horizontal Distribution

Let $M = G/T$, the G-flag manifold, and also let J denote a fixed integrable almost complex structure on M. Recall that there are $|W(G)|$ many such choices represented by the set of distinct Weyl chambers associated with the root system of G relative to T. Let Δ_+ denote the positive roots of the FWC determined by J. *Assume* that G is simple. Then there are exactly $r = \text{rank } G$ many simple roots (§6) which we denote by $\Delta_s = \{\varepsilon_s \theta_s : s \in I_s\} < \Delta_+$. Put $\Delta = \oplus\Sigma V_s, s \in I_s$. Then Δ is a $2r$-dimensional subspace of m, where $m = \mathcal{t}^\perp = \oplus\Sigma V_i, i \in I$. (Of course $I_s < I$.) Restricting $g_* \circ \pi_{*_e} : m \to T_T G/T \to T_{gT} G/T$ to Δ we obtain a $2r$-dimensional distribution, denoted by $\Delta(G/T)$, of G/T: to see that this distribution is well-defined just note that $\text{Ad}_t \Delta = \Delta$ for every $t \in T$.

Definition. ($G/T, J$ is in the above.)

$\Delta(G/T)$ is called the horizontal distribution of G/T.

Recall from §10 that J determines the forms $\Theta^i = \Omega^{2i-1} + \sqrt{-1}\varepsilon_i \Omega^{2i}$, $i \in I$, each up to $U(1)$-multiple. Consider the Pfaffian system Σ : $\{\Theta^a = 0, a \in I \backslash I_s\}$ on $M = G/T$ (we are pulling back forms to M, here.) Then Σ defines a C-rank r distribution in $TM^{(1,0)}$ ($TM^{(1,0)}$ is the so-called holomorphic tangent bundle of M) and we see that upon the identification $TM^{(1,0)} = TM$ this distribution is nothing but the horizontal distribution $\Delta(M)$.

We record that the exterior system Σ is in general overdetermined, although there do exist lower dimensional integrals.

Definition. An integral manifold of the system Σ (tangent spaces $< \Delta(G/T)$) is called a horizontal submanifold of G/T.

Example. $G = SU(n), T = S(U(1)^n) \cong U(1)^{n-1}$.

We use the notation introduced in §4. Choose FWC $= \{x = (x_i) :$ $x_1 > x_2 > \ldots > x_n\}$, $\Delta_+ = \{x_i - x_j : i < j\}$. The simple roots are $x_1 - x_2, x_2 - x_3, \ldots, x_{n-1} - x_n$. The "(1,0)-component" forms of Ω_m are $(\Omega_j^i), i < j$. The system Σ is given by $\{\omega_j^i = 0 : 1 \leq i < j \leq n, j \neq i + 1\}$, where (ω_j^i) are the pullbacks of (Ω_j^i). There are $n! - 1$ other choices of FWC resulting in completely similar discussions.

Chapter III

COMPLEX SUBMANIFOLDS

In this chapter we initiate the study of "pseudocomplex" maps into partial G-flag manifolds. Pseudocomplex maps into G/H are defined to be maps that have complex liftings into G/T. A partial G-flag manifold may not possess an invariant almost complex structure and hence complex manifold techniques are in general not available. For pseudocomplex maps, however, this problem is alleviated since we can study their complex liftings. Thus these maps form a natural object of our investigation.

§14 discusses general properties about complex maps into G/T: These properties are general in that the methods used to derive them do not directly depend upon the particular structure equations of G. We feel that this abstract approach to the study of complex maps into G/T may profitably be developed further. In §§15, 16 we deal with the cases $G = U(n), SO(2n; \mathbf{R})$ and examples are supplied.

§13. Pseudocomplex Maps

Let $M = G/H$ be a partial G-flag manifold. We will write $(G/H, J)$ to mean G/H equipped with a fixed invariant almost complex structure J.

Definition. Let S be any almost complex manifold. A smooth map $f : S \to (G/H, J)$ is said to be complex if $f_* TS^{(1,0)} < TM^{(1,0)}$, or equivalently $f_* \circ J_S = J \circ f_*$. We will also use the terms complex immersion (rank f_* is full everywhere) and complex submanifold interchangably.

45

Remark. Suppose that S and $(G/H, J)$ are both complex, that is, their respective almost complex structures are integrable. Then relative to any local holomorphic coordinates the local expression of $f : S \to (G/H, J)$ is holomorphic iff f is complex. (Easy exercise.) Recall the projection arising from the inclusion $H > T, \pi : G/T \to G/H$.

Definition. Let $f : S \to G/H$ be a smooth map. (Remember that G/H in general can not be given an invariant almost complex structure.) We will say that f is pseudocomplex if there exists an invariant almost complex structure J on G/H and a complex map $\hat{f} : S \to (G/T, J)$ such that $\pi \circ \hat{f} = f$. We now briefly discuss a theorem of Bryant which motivates the study of pseudocomplex maps. Roughly speaking, the theorem states that certain pseudocomplex maps are harmonic.

Let G/H be an irreducible type I inner symmetric space. It is well-known that in this case rank $G = $ rank H and that G/H is realized as a partial G-flag manifold. The complete list of such spaces is $SU(p + q)/S(U(p) \times U(q))$, $SO(p + q; \mathbf{R})/SO(p) \times SO(q), SO(2n; \mathbf{R})/U(n)$, $Sp(n)/U(n), Sp(p+q)/Sp(q) \times Sp(q)$ and all exceptional spaces except EI and EIV (cf. [He]).

Theorem. Consider the composition $\pi \circ g : S \to G/T \to G/H$, where S is a Riemann surface and G/H, an irreducible type I inner symmetric space. Suppose that g is horizontal (i.e. g_* is tangential to the horizontal distribution $\mathfrak{H}(G/T)$) and also suppose that g is holomorphic relative to an integrable invariant almost complex structure on G/T. Then $\pi \circ g$ is harmonic.

For the proof see [B3] §4.

In the case of an almost complex partial G-flag manifold we have

Proposition. Let $(G/H, J)$ be an almost complex partial G-flag manifold. Then there exists an invariant almost complex structure $\hat{J}$ on G/T so that the following holds: Let $g : S \to (G/T, \hat{J})$ be any complex map. Then $\pi \circ g : S \to (G/T, \hat{J}) \to (G/H, J)$ is also complex. Before proving the proposition we give a definition and a lemma.

Definition. Let $\hat{J}$ be an invariant almost complex structure on G/T and also let J be one on G/H. Identify $\hat{J}$ (respectively J) with an $\text{Ad}(T)$-invariant complex structure on $t^\perp$ (respectively $\text{Ad}(H)$-invariant complex structure on $h^\perp$). Note that $h^\perp$ is a subspace of $t^\perp$. We say that $\hat{J}$ restricts to J if $\hat{J}|_{h^\perp} = J$.

Lemma. Suppose $\hat{J}$ restricts to J and further suppose that $g : S \to (G/T, \hat{J})$ is a complex map. Then $\pi \circ g : S \to (G/T, \hat{J}) \to (G/H, J)$ is also complex.

Proof. Let J_S denote the given almost complex structure on S. g is complex, so $\hat{J} \circ g_* = g_* \circ J_S$. Also $\pi_* \circ \hat{J} = J \circ \pi_*$ since $\hat{J}$ restricts to J. Thus $J \circ \pi_* \circ g_* = \pi_* \circ \hat{J} \circ g_* = \pi_* \circ g_* \circ J_S$. This proves that $\pi \circ g$ is complex. q.e.d.

Proof of Proposition. J is the given $\text{Ad}(H)$-invariant complex structure on $h^\perp$. We want to find J on $t^\perp$ which restricts to J. Let $h^\perp = \oplus \Sigma n_\alpha$ be the irreducible decomposition of $h^\perp$ into its $\text{Ad}(H)$-invariant subspaces. Noting the containment $H > T$ we see that each n_α is a direct sum of $\text{Ad}(T)$-invariant subspaces V_i's. Recalling the discussion in §9 it is easy to see that we can choose an $\text{Ad}(T)$-invariant complex structure on $t^\perp$ so that its restriction to $h^\perp$ agrees with J. Note that we have no way of guaranteeing that the resulting almost complex structure on G/T is integrable. q.e.d.

§14. Moving Frames

Let $M^n = (G/H, J)$ be an almost complex partial G-flag manifold and also let S^p be any almost complex manifold. We consider a complex map $f : S \to M$.

Definition. Let U be an open set in S. A local lifting $e : U \to G$ of f ($\pi \circ e = f$) is called a moving frame or a G-frame along f.

Proposition. A smooth map $f : S \to M$ is complex if and only if $e^*\Omega_m^{(1,0)}$ is of type $(1,0)$ (i.e., $e^*(\Theta^\alpha)$ are all of type $(1,0)$), where e is any moving frame along f.

Proof. Recall that though the forms (Θ^α) are defined using an orthonormal basis of $h^\perp$ a change of orthonormal basis in $h^\perp$ will only change each Θ^α by a $U(1)$-multiple. Thus the type of $e^*(\Theta^\alpha)$ is well-defined. Now f is complex iff f^* pulls back type $(1,0)$ forms of M

to type $(1,0)$ forms of S. But $\pi^*(\pi : G \to M)$ establishes an isomorphism between type $(1,0)$ forms of M and the C-span of forms (Θ^α), and $\pi \circ e = f$. q.e.d.

Index Convention. $1 \leq i, j, k, \ldots \leq p, 1 \leq \alpha, \beta, \gamma, \ldots \leq n, p + 1 \leq a, b, c \ldots \leq n$.

Let $f : S \to M = (G/H, J)$ be a complex map. On S we pick a type $(1, 0)$ local coframe $\phi^1, \ldots, \phi^p$, written simply (ϕ^i). Let $E_1, \ldots, E_{2n}$ be an orthonormal basis of $m = h^\perp$ and write

$$\Omega_m^{(1,0)} = \Omega^{2\alpha-1} \otimes E_{2\alpha-1} + i\varepsilon_\alpha \Omega^{2\alpha} \otimes E_{2\alpha},$$
$$\Theta^\alpha = \Omega^{2\alpha-1} + i\varepsilon_\alpha \Omega^{2\alpha}.$$

Since f is complex, letting $e : U \to G$ be a moving frame along f we can find locally defined complex valued functions (Z_j^α) with

$$e^*\Theta^\alpha = Z_j^\alpha \phi^j.$$

$G \to M$ being a principal H-fibration, any other moving frame $\tilde{e} : \tilde{U} \to G$ along f is given by $\tilde{e} = e \cdot h$ on their common domain, where h is a smooth locally defined map into H. Recall the linear isotropy representation $i : H \to GL(m) = GL(2n; \mathbf{R})$ (using the reference basis $E_1, \ldots, E_{2n}$). Define the local functions $\tilde{Z} = (\tilde{Z}_j^\alpha)$ by $\tilde{e}^*\Theta^\alpha = \tilde{Z}_j^\alpha \phi^j$. Then we get the transformation rule $\tilde{Z} = i(h^{-1})Z$, where we think of Z and $\tilde{Z}$ as $2n$ by p real matrices in an obvious way, $(Z_j^\alpha) \leftrightarrow (\operatorname{Re} Z_j^\alpha, \operatorname{Im} Z_j^\alpha)$ etc. The following observations are immediate.

Observations.

i) The rank of $Z = (Z_j^\alpha)$ is an invariant. That is to say, it is independent of the moving frame chosen. Hence it defines a global integer valued function on S.

ii) $f : S \to M$ is an immersion iff rank Z is full $(= p)$ everywhere.

In the following we specialize our discussion to the case of $M = G/T$. Recall the orthogonal decomposition $g = t \oplus \Sigma V_\alpha, m = t^\perp = \oplus \Sigma V_\alpha$. Choose an orthonormal basis $\{E_1, \ldots, E_{2n}\}$ of m and order it so that R-span $\{E_1, E_2\} = V_1$, etc. Upon identifying $T_0 M$ with m the linear isotropy representation $i : T \to GL(m)$ becomes the adjoint representation Ad: $T \to GL(m)$. More specifically, $\mathrm{Ad}|_{V_\alpha}$ is given by the matrix relative to the basis $\{E_{2\alpha-1}, E_{2\alpha}\}$

$$t \mapsto \begin{pmatrix} \cos 2\pi\Theta_\alpha(t) & -\sin 2\pi\Theta_\alpha(t) \\ \sin 2\pi\Theta_\alpha(t) & \cos 2\pi\Theta_\alpha(t) \end{pmatrix},$$

where Θ_α is the α-th root of G. Hence we obtain the following matrix representation for i relative to the reference basis $\{E_1, \ldots, E_{2n}\}$:

$$t \mapsto \mathrm{diag}\left(\begin{pmatrix} \cos 2\pi\Theta_1(t) & -s \\ s & c \end{pmatrix}, \ldots, \begin{pmatrix} \cos 2\pi\Theta_n(t) & -s \\ s & c \end{pmatrix} \right)$$

in $SO(2)^n < GL(2n; \mathbf{R})$, where $s = \sin 2\pi\Theta_\alpha(t)$, etc. Using the standard identification $U(1) = SO(2)$ and rescaling the roots we write

$$i : t \mapsto \mathrm{diag}\left(e^{i\Theta_1(t)}, \ldots, e^{i\Theta_n(t)}\right).$$

Let $f : S^p \to M = (G/T, J)$ be a complex map. As in the above discussion we have $e^*\Theta^\alpha = Z_i^\alpha \phi^i$, where (ϕ^i) is a type $(1,0)$ local coframe on S, e is a G-frame along f and (Z_i^α) are some complex valued local functions on S.

Let $\tilde{e}$ be another moving frame given by $\tilde{e} = e \cdot t$ with t, a smooth map into T. Write $t = (e^{it_1}, \ldots, e^{it_n})$, where (t_α) are real valued local functions on S. Define $\tilde{Z} = (\tilde{Z}_i^\alpha)$ by $\tilde{e}^*\Theta^\alpha = \tilde{Z}_i^\alpha \phi^i$. Then from the formulae $\tilde{Z} = i(t^{-1})Z$ and $i(t^{-1}) = \mathrm{diag}(e^{-i\Theta_1(t)}, \ldots, e^{-i\Theta_n(t)})$ we get

$$\tilde{Z}_i^\alpha = e^{-\sqrt{-1}\Theta_\alpha(t)} Z_i^\alpha,$$

where Θ_α's are the rescaled roots of G. This means that the nonzero rows of the matrix Z define points in S^{2p-1}. The number of points in S^{2p-1} defined by the nonzero rows of Z is obviously an invariant, i.e., independent of the G-frame chosen to express Z. This number then defines a global integer valued function $\# : S \to Z^+$. Also define $r_i^\alpha = Z_i^\alpha \tilde{Z}_i^\alpha$. Then $r_i^\alpha = \tilde{r}_i^\alpha$ and each defines a global function $r_i^\alpha : S \to R \geq 0$.

Remark. Using the Maurer-Cartan structure equations $d\Omega = -\Omega \wedge \Omega$ of G together with the fact that d commutes with $*$ one can differentiate the equations $e^*\Theta^\alpha = Z_i^\alpha \phi^i$ and obtain the so-called local structure equations of the map f.

§15. $U(n)$-flag Manifold

Let $G = U(n)$ and $T = U(1) \times \ldots \times U(1) = U(1)^n$. Then $G/T = F_{1,2,\ldots,n}(\mathbb{C}^n)$ = the full complex flag manifold. (We could use $G = SU(n)$ and $T = S(U(1)^n)$. The choice is mostly a matter of taste here.) We fix an invariant almost complex structure J on G/T.

Recall the decomposition from §12, $g = t \oplus \Sigma V_{\alpha\beta}(1 \leq \alpha < \beta \leq n)$, where $V_{\alpha\beta} = R$-span $\{E_{\alpha\beta}, F_{\alpha\beta}\}$.

The component forms in $\Omega_{t^\perp}^{(1,0)}$ are $(\Theta^{\alpha\beta})$ and $\Theta^{\alpha\beta}$ is equal to Ω_β^α if $\varepsilon_{\alpha\beta} = +1$ and is equal to $\bar{\Omega}_\beta^\alpha$ if $\varepsilon_{\alpha\beta} = -1$.

We compute the matrix isotropy representation $i : U(1)^n \to GL(t^\perp) = GL(n^2 - n; R)$ relative to the basis $\{E_{\alpha\beta}, F_{\alpha\beta}\}$. Let $t = (e^{it_1}, \ldots, e^{it_n}) \in U(1)^n = T$. We saw earlier that

$$\mathrm{Ad}|_{V_{\alpha\beta}} : t \mapsto \begin{pmatrix} \cos(t_\alpha - t_\beta) & -\sin(t_\alpha - t_\beta) \\ \sin(t_\alpha - t_\beta) & \cos(t_\alpha - t_\beta) \end{pmatrix}.$$

Thus with the usual identification $SO(2) = U(1)$ we write $i(t) = \mathrm{diag}(e^{i(t_1-t_2)}, e^{i(t_2-t_3)}, \ldots, e^{i(t_{n-1}-t_n)}) \in U(\frac{1}{2}(n^2 - n)) < SO(n^2 - n; R)$.

Let $f : S^p \to G/T$ be a complex map. Choose a type $(1,0)$ local coframe (ϕ^i) on S. Choosing a $U(n)$-frame e along f we write $e^*\Theta^{\alpha\beta} = Z_i^{\alpha\beta}\phi^i$, $1 \leq i \leq p$, $1 \leq \alpha < \beta \leq n$, and $(Z_i^{\alpha\beta})$ complex valued local functions on S. Imposing the lexicographical ordering on the double indices $(\alpha\beta)$ we may rewrite the above equations as

$$e^*\Theta^A = Z_i^A \phi^i, \quad 1 \leq A \leq \frac{1}{2}(n^2 - n).$$

Let $\tilde{e} = e \cdot t, t = (e^{it_1}, \ldots, e^{it_n}): U \cap \tilde{U} \to T$ be another $U(n)$-frame and define the matrix $\tilde{Z}$ by $\tilde{e}^*\Theta^A = \tilde{Z}_i^A \phi^i$. Then Z and $\tilde{Z}$ are related by the formula

$$\tilde{Z} = \mathrm{diag}(e^{i(t_2-t_1)}, e^{i(t_3-t_2)}, \ldots, e^{i(t_n-t_{n-1})})Z,$$

or more compactly,

$$\tilde{Z}_i^{\alpha\beta} = e^{i(t_\beta - t_\alpha)} Z_i^{\alpha\beta}.$$

It follows that $r_i^{\alpha\beta} = Z_i^{\alpha\beta} \bar{Z}_i^{\alpha\beta}$ are globally defined invariants of f.

We will compute the structure equations for complex curves, so $p = 1$. Let us fix a Riemannian metric on S in the conformal class of its complex structure and let ϕ denote a type $(1, 0)$ local unitary coframe on S.

Notation. $e^* \Omega^\alpha_\beta = \omega^\alpha_\beta, e^* \Theta^{\alpha\beta} = \theta^{\alpha\beta}$.
There are the equations

$$d\omega^\alpha_\beta = -\omega^\alpha_\gamma \wedge \omega^\gamma_\beta,$$

$$d\phi = i\omega \wedge \phi,$$

where ω is the Levi-Civita connection form of (S, ds^2) relative to ϕ.

Consulting the above equations we differentiate the both sides of $\theta^{\alpha\beta} = Z^{\alpha\beta}\phi$. For the sake of simplicity we assume that $\theta^{\alpha\beta} = \omega^\alpha_\beta, 1 \leq \alpha < \beta \leq n$. We then obtain

$$(-\bar{Z}^\gamma_\alpha Z^\gamma_\beta + Z^\alpha_\delta \bar{Z}^\beta_\delta)\phi \wedge \bar{\phi} - (\omega^\alpha_\alpha - \omega^\beta_\beta) \wedge Z^\alpha_\beta \phi = (dZ^\alpha_\beta + iZ^\alpha_\beta \omega) \wedge \phi,$$

where $Z^\alpha_\beta = Z^{\alpha\beta}, \gamma < \alpha, \delta > \beta$ and sum on δ and γ.

Looking at the above equations we see that they would yield relationships among $\omega, \omega^\alpha_\alpha - \omega^\beta_\beta$ and the forms involving the components of Z and their derivatives. Differentiating these resulting relationships one obtains equations relating the Gaussian curvature of (S, ds^2) and certain extrinsic invariants derived from $r^{\alpha\beta}_j$.

Remarks.

i) In the next chapter we will carry out the above indicated calculations in specific instances thereby "classifying" classes of maps into G/T.

ii) From the viewpoint of a moving frame theorist the analysis of complex maps into $U(n)/U(1)^n$ (or indeed into G/T in general) is relatively easy in that it does not require any "higher order reduction" of the frame bundle $f^{-1}G \to S$. This is precisely why it is advantageous to approach the study of certain maps into more general spaces of the form G/H via projections and liftings from and into G/T.

We now look at some partial $U(n)$-flag manifolds.

If one insists on having a *complex* partial $U(n)$-flag manifold $U(n)/H$ then Wang's theorem in §9 says that H must be the centralizer of a torus in $U(n)$ (and conversely such a space is complex). The following lemma is straightforward.

Lemma. $H < U(n)$ is the centralizer of a torus in $U(n)$ if and only if $H = U(n_1) \times \ldots \times U(n_k)$ for some positive integers (n_i) with $\Sigma n_i = n$.

In the above case we obtain $U(n)/H = F_{n_1,\ldots,n_k}(\mathbb{C}^n)$ which is a manifold of partial complex flags. Most special among them are the two hermitian symmetric cases $U(n)/U(1) \times U(n-1) = CP^{n-1}$, the complex projective space and more generally $U(n)/U(n_1) \times U(n - n_1) = G_{n_1}(\mathbb{C}^n)$, the complex Grassmannian.

Example. $G = U(4), H = U(2) \times U(1) \times U(1)$,

$G/H = F_{2,3,4}(C^4)$. This is not a symmetric space. (Incidentally one may recall that earlier we have uncovered at least two integrable invariant almost complex structures on this space which are not related by a diffeomorphism and this is of course impossible for a symmetric space.) There is the orthogonal decomposition $g = h \oplus m$. With the usual notation $m = R$-span $\{E_{13}, F_{13}, E_{14}, F_{14}, E_{23}, F_{23}, E_{24}, F_{24}, E_{34}, F_{34}\}$. Using the complex notation we represent a vector in m as $v = \Sigma x_{\alpha\beta} E_{\alpha\beta} + i y_{\alpha\beta} F_{\alpha\beta}$. Putting $z_{\alpha\beta} = x_{\alpha\beta} + i y_{\alpha\beta}$ and reindexing lexicographically we represent v as the matrix given by

$$v = \left[\begin{array}{c|cc} 0 & z_1, z_2 \\ & z_3, z_4 \\ \hline & 0 \;\; z_5 \\ * & * \;\; 0 \end{array} \right], \quad \begin{array}{l} \text{where } * = -\text{conjugate transpose,} \\ z_1 = z_{13} \text{ etc.} \end{array}$$

The linear isotropy representation $i : U(2) \times U(1) \times U(1) \to GL(10; R)$ relative to the basis $\{E_{\alpha\beta}, F_{\alpha\beta}\}$ is computed using the formula $\mathrm{Ad}_h v = hvh^{-1}, h \in H$. If $h = (A, e^{it_1}, e^{it_2}), A \in U(2)$ then

$$i : \begin{pmatrix} z_1 & z_2 \\ z_3 & z_4 \end{pmatrix} \mapsto A \begin{pmatrix} z_1 & z_2 \\ z_3 & z_4 \end{pmatrix} \begin{pmatrix} e^{-it_1} & 0 \\ 0 & e^{-it_2} \end{pmatrix},$$

$$z_5 \mapsto e^{i(t_1 - t_2)} z_5.$$

We see that the isotropy action of H on the matrix $Z = (Z_i^{\alpha\beta})$ of a complex map $f : S \to G/H$ is rather complicated and invariants not as easily computed.

§16. $SO(2n; \mathbf{R})$-flag Manifolds

Let $G = SO(2n; \mathbf{R}), T = SO(2)^n$. Then $G/T = \tilde{F}_{2,4,\ldots,2n}(\mathbf{R}^{2n})$, the full oriented even flags in $\mathbf{R}^{2n}$. We fix an invariant almost complex structure J on G/T.

Note that the standard injection $U(n) \to SO(2n; \mathbf{R})$ projects down to a homogeneous imbedding (indeed totally geodesic) $U(n)/U(1)^n \to SO(2n; \mathbf{R})/SO(2)^n = SO(2n; \mathbf{R})/U(1)^n$. Thus the case at hand considerably generalizes the previous case of §15.

Recall from §12 the orthogonal decomposition $g = t \oplus t^\perp$. $t^\perp = \oplus \Sigma V_{\alpha\beta} \oplus \Sigma V'_{\alpha\beta} (1 \leq \alpha < \beta \leq n)$, where $V_{\alpha\beta} = R$-span $\{E_{\alpha\beta}, F_{\alpha\beta}\}$, $V'_{\alpha\beta} = R$-span $\{E'_{\alpha\beta}, F'_{\alpha\beta}\}$.

We compute the matrix isotropy representation $i : SO(2)^n \to GL(t^\perp) = GL(2n^2 - 2n; \mathbf{R})$ relative to the basis $\{E_{\alpha\beta}, F_{\alpha\beta}, E'_{\alpha\beta}, F'_{\alpha\beta}\}$. Via the usual identification $SO(2) = U(1)$ we let $t = \text{diag}(e^{it_1}, \ldots, e^{it_n}) \in T$. Put $z_{\alpha\beta} = xE_{\alpha\beta} + iyF_{\alpha\beta}$ and $z'_{\alpha\beta} = xE'_{\alpha\beta} + iyF'_{\alpha\beta}, x, y \in \mathbf{R}$. Then $i(t) : z_{\alpha\beta} \mapsto e^{i(t_\alpha - t_\beta)} z_{\alpha\beta}$ and $z'_{\alpha\beta} \mapsto e^{i(t_\alpha + t_\beta)} z'_{\alpha\beta}$. Putting them together

$$i : t \mapsto \text{diag}(e^{i(t_1 - t_2)}, \ldots, e^{i(t_{n-1} - t_n)}, e^{i(t_1 + t_2)}, \ldots, e^{i(t_{n-1} + t_n)})$$

in $U(n^2 - n) < SO(2n^2 - 2n; \mathbf{R})$.

Let $f : S^p \to G/T$ be a complex map. On S we pick a type $(1,0)$ local coframe, say $\phi^1, \ldots, \phi^p$. Now the "$(1,0)$-component" forms of $\Omega_{t^\perp}$ are given by

$$\Theta^{\alpha\beta} = \frac{1}{2}[(\Omega_{2\beta-1}^{2\alpha-1} + \Omega_{2\beta}^{2\alpha}) + i\varepsilon_{\alpha\beta}(\Omega_{2\beta-1}^{2\alpha} - \Omega_{2\beta}^{2\alpha-1})],$$

$$\Theta'^{\alpha\beta} = \frac{1}{2}[(\Omega_{2\beta-1}^{2\alpha-1} - \Omega_{2\beta}^{2\alpha}) + i\varepsilon_{\alpha\beta}(\Omega_{2\beta-1}^{2\alpha} + \Omega_{2\beta}^{2\alpha-1})],$$

where $\varepsilon_{\alpha\beta} = +1$ or -1, the sign dictated by J.

Choosing a moving frame e along f we write

$$e^* \Theta^{\alpha\beta} = Z_i^{\alpha\beta} \phi^i, \quad e^* \Theta'^{\alpha\beta} = Z_i'^{\alpha\beta} \phi^i$$

for some complex valued local functions Z, Z' on S.

We now compute the structure equations for complex curves. So $p = 1$.

Notation. $e^* \Theta^{\alpha\beta} = \theta^{\alpha\beta}, e^* \Omega_B^A = \omega_B^A$, etc.

Let ω denote the connection form on S relative to ϕ, i.e., $d\phi = i\omega \wedge \phi$. Using the structure equations $d\Omega = -\Omega \wedge \Omega$ we exterior differentiate both sides of the equations $\theta^{\alpha\beta} = Z^{\alpha\beta}\phi$, $\theta'^{\alpha\beta} = Z'^{\alpha\beta}\phi$. This leads to the (local) structure equations for f:

$$- \sum_{\gamma > \alpha} [(Z^{\alpha\gamma} + Z'^{\alpha\gamma})\phi \wedge \omega_{2\beta-1}^{2\gamma-1} + i(Z^{\alpha\gamma} - Z'^{\alpha\gamma})\phi \wedge \omega_{2\beta-1}^{2\gamma}]$$

$$+ \sum_{\delta < \alpha} [(\phi_{\delta\alpha}^1 + i\phi_{\delta\alpha}^2) \wedge \omega_{2\beta-1}^{2\alpha-1} + (\phi_{\delta\alpha}^3 + i\phi_{\delta\alpha}^4) \wedge \omega_{2\beta-1}^{2\delta}] +$$

$$i\omega_{2\alpha-1}^{2\alpha} \wedge (Z^{\alpha\beta} + Z'^{\alpha\beta})\phi = [d(Z^{\alpha\beta} + Z'^{\alpha\beta}) + i(z^{\alpha\beta} + Z'^{\alpha\beta})\omega] \wedge \phi$$

and

$$- \sum_{\gamma > \alpha} [(Z^{\alpha\gamma} + Z'^{\alpha\gamma})\phi \wedge \omega_{2\beta}^{2\gamma-1} + i(Z^{\alpha\gamma} - Z'^{\alpha\gamma})\phi \wedge \omega_{2\beta}^{2\gamma}]$$

$$+ \sum_{\delta < \alpha} [(\phi_{\delta\alpha}^1 + i\phi_{\delta\alpha}^2) \wedge \omega_{2\beta}^{2\delta-1} + (\phi_{\delta\alpha}^3 + i\phi_{\delta\alpha}^4) \wedge \omega_{2\beta}^{2\delta}] +$$

$$\omega_{2\alpha-1}^{2\alpha} \wedge (Z^{\alpha\beta} - Z'^{\alpha\beta})\phi = [id(Z^{\alpha\beta} - Z'^{\alpha\beta}) - (Z^{\alpha\beta} - Z'^{\alpha\beta})\omega] \wedge \phi,$$

where $\phi_{\delta\alpha}^1, \phi_{\delta\alpha}^2, \phi_{\delta\alpha}^3, \phi_{\delta\alpha}^4$ are real 1-forms defined by

$$(Z^{\delta\alpha} + Z'^{\delta\alpha})\phi = \phi_{\delta\alpha}^1 + i\phi_{\delta\alpha}^3, \quad i(Z^{\delta\alpha} - Z'^{\delta\alpha})\phi = \phi_{\delta\alpha}^2 + i\phi_{\delta\alpha}^4.$$

If we define $\tilde{Z} = (\tilde{Z}_i^{\alpha\beta})$ by the equations $\tilde{e}^* \Theta^{\alpha\beta} = \tilde{Z}_i^{\alpha\beta}\phi^i$ and likewise define $\tilde{Z}'$ with $\tilde{e} = e \cdot t$ as usual then the transformation rule is computed, from the matrix isotropy representation, to be

$$\tilde{Z}_i^{\alpha\beta} = e^{\sqrt{-1}(t_\beta - t_\alpha)} Z_i^{\alpha\beta}, \quad \tilde{Z}_i'^{\alpha\beta} = e^{-\sqrt{-1}(t_\alpha + t_\beta)} Z_i'^{\alpha\beta}.$$

We thus obtain the globally defined invariants $r_i^{\alpha\beta} = Z_i^{\alpha\beta} \bar{Z}_i^{\alpha\beta}$ and $r_i'^{\alpha\beta} = Z_i'^{\alpha\beta} \bar{Z}_i'^{\alpha\beta}$.

Among the partial $SO(m)$-flag manifolds we have the two hermitian symmetric cases, $SO(n+2; \mathbf{R})/SO(2) \times SO(n)$, $SO(2n; \mathbf{R})/U(n)$.

$SO(n + 2; \mathbf{R})/SO(2) \times SO(n)$ may be realized as the nondegenerate hyperquadric Q_n in CP^{n+1}. Q_n contains as its maximal linear subspaces $CP^{[n/2]}$, hence one should consider the study of complex maps into the quadric to be more general in scope than that of complex maps into the complex projective space.

Example. $G = SO(6; \mathbf{R}), H = SO(2) \times SO(4)$.

$G/T = Q_4$. We have $g = h \oplus m$, where $m = R$-span $\{E_{12}, F_{12}, E_{13}, F_{13}, E'_{12}, F'_{12}, E'_{13}, F'_{13}\}$. The "(1,0)-component" forms of Ω_m are

$$\Theta^{12} = \frac{1}{2}[(\Omega_3^1 + \Omega_4^2) + i\varepsilon_{12}(\Omega_3^2 - \Omega_4^1)],$$

$$\Theta^{13} = \frac{1}{2}[(\Omega_5^1 + \Omega_6^2) + i\varepsilon_{13}(\Omega_5^2 - \Omega_6^1)],$$

$$\Theta'^{12} = \frac{1}{2}[(\Omega_3^1 - \Omega_4^2) + i\varepsilon_{12}(\Omega_3^2 + \Omega_4^1)],$$

$$\Theta'^{13} = \frac{1}{2}[(\Omega_5^1 - \Omega_6^2) + i\varepsilon_{13}(\Omega_5^2 + \Omega_6^1)],$$

It follows that the pullbacks of the forms $\Omega_1^a + i\varepsilon_a \Omega_2^a, \varepsilon_a^2 = 1, 3 \leq a \leq 6$, are type $(1,0)$ on Q_4. The standard metric is given by the pullback of

$$\Theta^{12}\bar{\Theta}^{12} + \ldots + \Theta'^{13}\bar{\Theta}'^{13} = \frac{1}{2}\Sigma(\Omega_i^a)^2, \quad i = 1, 2, \ 3 \leq a \leq 6.$$

We compute the isotropy representation $i : SO(2) \times SO(4) \to GL(m) = GL(8; \mathbf{R})$. Let $t = (e^{it}, A)$ be an element of $U(1) \times SO(4) = SO(2) \times SO(4)$ and $X \in m$. (Here we have an obvious notational abuse.) From the matrix multiplication

$$\begin{bmatrix} e^{it} & 0 \\ 0 & A \end{bmatrix} \begin{bmatrix} 0 & X \\ -{}^t X & 0 \end{bmatrix} \begin{bmatrix} e^{-it} & 0 \\ 0 & A^{-1} \end{bmatrix}$$

we see that

$$i(t) : X \mapsto \begin{pmatrix} \cos t & -\sin t \\ \sin t & \cos t \end{pmatrix} X A^{-1}.$$

The above action is sufficiently complicated that even in the study of complex curves in Q_4 there seems to be some formidable difficulties pertaining to moving frame theoretic analysis. See [J-R-Y].

Remarks.

i) Considering the symplectic case $Sp(n)/U(1)^n$ ($U(1)^n = T$, a maximal torus in $Sp(n)$) we note that the natural injection $U(1) \hookrightarrow Sp(1)$ induces the projection $Sp(n)/U(1)^n \to Sp(1)/Sp(1) \times Sp(n-1) = HP^{n-1}$, the quaternionic projective space, or more generally the projection $Sp(n)/U(1)^n \to Sp(n)/Sp(p) \times Sp(n-p) =$ the quaternionic Grassmannian.

ii) Going to an exceptional case consider G_2/T. Then the standard 6-sphere may be realized as a partial G_2-flag manifold (almost complex, never integrable) $G_2/SU(3)$. See [B1].

Chapter IV

EXAMPLES: HOLOMORPHIC CURVES

Holomorphic curves in various spaces are investigated here. Our goal throughout the chapter has been an understanding of global properties. With this in mind several integral formulae are derived. These formulae should be thought of as a sort of Plücker relations.

§17 begins a study of complex curves in the $U(n)$-flag manifold. An obstruction to further understanding of this case is uncovered and discussed. It has to do with a possible behavior of the zeros of invariants.

§§18 and 19 go together in that horizontal curves in $U(m+1)/T$ "are" just holomorphic curves in CP^m. For a precise statement of this look at the theorem in §19 and also the discussion preceding it.

§20 gives a description of horizontal curves in $Sp(n)/T$ and these curves are seen to generalize horizontal curves in $U(n)/T$. (See the remark following the discussion of the degenerate case.)

In §§18–20 we may claim to have given complete descriptions of the curves involved. In each case, to a curve $f : S \rightarrow G/T$ we associate a sort of Gauss map, called the Gauss-Frenet map of f, $\Phi_f : S \rightarrow S^{r-1} <$ $\mathbf{R}^r$, where $r = \operatorname{rank} G$. (The map Φ_f in turn completely determines f.) The component functions of Φ_f satisfy a system of generalized Euler-Lagrange equations. In the language of exterior differential systems these equations are the complete integrability conditions. To put it another way, given functions (a_i) satisfying a system of generalized Euler-Lagrange equations, one can produce f with $\Phi_f = (a_i)$ using integration involving ordinary differential equations only.

For a treatment of holomorphic curves in $SO(n; \mathbf{R})$-flag manifolds the reader is refered to [J-R-Y] and [Y]. We mention that horizontal curves in the $SO(n; \mathbb{R})$-flag manifold correspond to "isotropic" holomorphic curves in the complex hyperquadric. These curves are closely related to minimal surfaces in spheres and in real projective spaces.

§17. Holomorphic Curves in $F_{1,2,3}(\mathbb{C}^3)$

Let $M = F_{1,2,\dots,n}(\mathbb{C}^n) = SU(n)/S(U(1)^n) = U(n)/U(1)^n$. We will use $G = U(n)$ and $T = U(1)^n$. $\Omega = (\Omega_\beta^\alpha), 1 \leq \alpha, \beta \leq n$, denotes the $\mathcal{U}(n)$-valued Maurer-Cartan form of G as usual. Pick J with $\Theta^{\alpha\beta} = \Omega_\beta^\alpha, 1 \leq \alpha < \beta \leq n$. Note that J is integrable.

Let $f : S \to M$ be a holomorphic immersion, where S is a connected Riemann surface.

Remark. Our interest here is primarily differential-geometric as opposed to algebraic-geometric. In view of this the immersion assumption on f seems adequate. At any rate the possible branch points are isolated since f is holomorphic.

Given a (local) moving frame e along f we define local complex valued functions $(Z_\beta^\alpha), 1 \leq \alpha < \beta \leq n$, on S by

$$e^* \Omega_\beta^\alpha = Z_\beta^\alpha \phi,$$

where ϕ is a type $(1,0)$ local coframe on S.

If $\tilde{e} = e \cdot t, t = (e^{it_\alpha}) = U(1)^n$-valued local function on S, is another moving frame along f, then as we saw in §15

$$\tilde{Z}_\beta^\alpha = e^{i(t_\beta - t_\alpha)} Z_\beta^\alpha, \quad 1 \leq \alpha < \beta \leq n.$$

Definition. Let $f : S \to M$ be a holomorphic immersion and $p \in S$. Then p is called a regular point of the map f if $\prod Z_\beta^\alpha(p) \neq 0 (1 \leq \alpha < \beta \leq n)$. Otherwise p is called a singular point of f.

The notion of regularity (or singularity) of a point is well-defined since the zero set of each Z_β^α does not depend on the particular moving frame chosen to express Z_β^α. Note that $Z_\beta^\alpha(p) = 0$ for every α, β iff p is a branch point $(f_{*p} = 0)$. So for immersions this can never happen.

Proposition. Let $n = 3$, so $M = F_{1,2,3}(\mathbb{C}^3)$. Then in a neighborhood of any regular point of $f : S \to M$ one can choose a moving frame e so that relative to $e, (Z_\beta^\alpha)$ are all positive real valued. This is in general not possible for $n > 3$.

Proof. The problem of making (Z_β^α) simultaneously positive and real boils down to solving a linear system with $(n-1)n/2$ equations in n unknowns. In a neighborhood of a regular point in which all points are regular (existence of which is guaranteed by continuity) this linear system admits a unique solution if $n = 3$, and is overdetermined if $n > 3$. q.e.d.

Notation. $e^* \Omega_\beta^\alpha = \omega_\beta^\alpha$.

For the rest of this section we work exclusively with the case $n = 3$. Choose a moving frame e, whenever possible (certainly near a regular point), that makes (Z_β^α) positive real valued and write

$$\omega_2^1 = a\phi, \quad \omega_3^1 = b\phi, \quad \omega_3^2 = c\phi, \quad a, b, c > 0.$$

In general, in a neighborhood of an arbitrary point we have to allow $\omega_2^1 = Z_2^1 \phi, \omega_3^1 = Z_3^1 \phi, \omega_3^2 = Z_3^2 \phi$ with (Z_β^α) complex valued.

Let ω denote the (real) connection form on S relative to ϕ so that $d\phi = i\omega \wedge \phi$. Using the Maurer-Cartan structure equations $d\Omega = -\Omega \wedge \Omega$ of G, exterior differentiation on both sides of the above equations yields the following structure equations for f: Near a regular point,

$$[bc\bar{\phi} + da + ia\omega + a(\omega_1^1 - \omega_2^2)] \wedge \phi = 0,$$
$$[db + ib\omega + b(\omega_1^1 - \omega_3^3)] \wedge \phi = 0,$$
$$[-ab\bar{\phi} + dc + ic\omega + c(\omega_2^2 - \omega_3^3)] \wedge \phi = 0.$$

Near a general point,

$$[Z_3^1 Z_3^2 \bar{\phi} + dZ_2^1 + iZ_2^1 \omega + Z_2^1(\omega_1^1 - \omega_2^2)] \wedge \phi = 0,$$
$$[dZ_3^1 + iZ_3^1 \omega + Z_3^1(\omega_1^1 - \omega_3^3)] \wedge \phi = 0, \qquad (*)$$
$$[-Z_2^1 Z_3^1 \bar{\phi} + dZ_3^2 + iZ_3^2 \omega + Z_3^2(\omega_2^2 - \omega_3^3)] \wedge \phi = 0.$$

Recall (§15) that the functions $r_2^1 = Z_2^1 \bar{Z}_2^1, r_3^1 = Z_3^1 \bar{Z}_3^1, r_3^2 = Z_3^2 \bar{Z}_3^2$ are globally defined invariants on S.

Proposition. r_3^1 is either identically zero or its zeros are isolated.

Proof. Assume that r_3^1 is not identically zero. Rewriting the equation (*) we get $dZ_3^1 = -Z_3^1(i\omega + \omega_1^1 - \omega_3^3)(\mod \phi)$. It now follows from a theorem of Chern ([Ch1] §4) that the zeros of Z_3^1 are isolated. Moreover, the theorem states that the zeros are all of finite multiplicity. Now the zero set of Z_3^1 (which is globally defined) coincides with the zero set of r_3^1. q.e.d.

Remark. In contrast to the above proposition the behavior of the zeros of r_2^1 and r_3^2 are generally not predictable. This difficulty seems to be responsible for the lack of nice global theory of holomorphic curves under investigation.

Functions whose zeros behave as in the above proof occur frequently in our study and this motivates the following definition.

Definition. Let U be a domain in S. A complex valued smooth function $h : U \to \mathbf{C}$ is said to be of analytic type if for each point $x \in U$, if z is a local holomorphic coordinate centered at x, then

$$h = z^b \tilde{h},$$

where b is a positive integer and $\tilde{h}$ is a smooth complex valued function with $\tilde{h}(x) \neq 0$.

It is known that the functions of analytic type are exactly solutions of exterior equation

$$dh = h\psi(\mod \phi),$$

where ψ is a complex valued 1-form on U and ϕ is a nowhere zero type $(1,0)$ form on U. So if h is a function of analytic type on U, then h is either identically zero on U or its zeros are isolated and of finite multiplicity (the integer b in the above definition is the multiplicity at x).

Assume that r_3^1 is not identically zero. (The case $r_3^1 \equiv 0$ will follow.) Then the equation $\omega_3^1 = b\phi$ is valid near every point of S except at an isolated set where r_3^1 is zero.

We equip M with a $U(3)$-invariant metric given by the pullback of $\Sigma \Omega_\beta^\alpha \bar{\Omega}_\beta^\alpha (1 \leq \alpha < \beta \leq 3)$. Assign S the metric induced by f so that

$f : S \to M$ becomes an isometric immersion. We further assume that ϕ is unitary so that $r_2^1 + r_3^1 + r_3^2 = 1$.

Exterior differentiation of both sizes of the equation $\omega_3^1 = b\phi$ leads to

$$\omega - i\left(\omega_I^1 - \omega_3^3\right) = *d\log b\,,$$

where ω is the Levi-Civita connection form relative to ϕ and $*$ is the Hodge operator of S, ds^2.

Note that $\log b$ is defined everywhere except at the isolated zero set of r_3^1 at which it goes to $-\infty$.

Let K denote the Gaussian curvature of S, ds^2. Then we have $d\omega = \frac{i}{2} K \phi \wedge \bar{\phi}$. Also let Δ denote the Laplace-Beltrami operator of (S, ds^2) so that $d^*d\log b = \frac{i}{2}(\Delta \log b)\phi \wedge \bar{\phi}$.

Another differentiation now yields

$$K - 2\left(1 + b^2\right) = \Delta \log b\,.$$

Assume now that S is compact (without boundary.) S, already being a connected Riemann surface, is homeomorphic to $T_g = $ the torus with g handles. The Euler-Poincaré characteristic χ_S of S is equal to $2 - 2g$. We have

$$\text{area}(S) = \frac{i}{2} \int_S \phi \wedge \bar{\phi}\,,$$

$$\chi_S = \frac{i}{4\pi} \int_S K \phi \wedge \bar{\phi}\,,$$

$$\#\left(r_3^1\right) = -\frac{i}{2\pi} \int_S \Delta \log b\phi \wedge \bar{\phi}\,,$$

where $\#\left(r_3^1\right)$ denotes the number of zeros of r_3^1 each counted with multiplicity. The second equality is the Gauss-Bonnet-Chern theorem and the last equality is an elementary application of the argument principle.

The following integral formula now follows easily.

$$2\chi_S + \#\left(r_3^1\right) = \frac{2}{\pi}\text{area}(S) + \frac{i}{\pi} \int_S r_3^1 \phi \wedge \bar{\phi}\,.$$

We turn now to the case r_3^1 identically zero. We have

Proposition. Horizontal holomorphic curves in (M, J) are precisely those holomorphic curves with r_3^1 identically zero.

Proof. $M = F_{1,2,3}(\mathbf{C}^3) = SU(3)/S(U(1)^3)$. The simple roots of $SU(3)$ corresponding to J (remember that J goes with the choice $\Theta^{\alpha\beta} = \Omega_\beta^\alpha, 1 \leq \alpha < \beta \leq 3$) are $x_1 - x_2$ and $x_2 - x_3$. (See §12.) Thus the associated system Σ on M defining the horizontal distribution is given by $\{\omega_3^1 = 0\}$. The rest is easy. q.e.d.

§18. Horizontal Curves in $U(n)/T$

Let $G = U(n), T = U(1)^n$. Then $M = G/T = F_{1,2,\dots,n}(\mathbf{C}^n)$, the full complex flag manifold. $\Omega = (\Omega_\beta^\alpha), 1 \leq \alpha, \beta \leq n$, is the $\mathcal{U}(n)$-valued Maurer-Cartan form of G and we pick J with $\Theta^{\alpha\beta} = \Omega_\beta^\alpha, 1 \leq \alpha < \beta \leq n$, as in §17.

J determines the fundamental Weyl chamber given by $\{x = (x_\alpha) : x_1 > x_2 > \dots > x_n\}$ and the positive roots $\Delta_+ = \{x_\alpha - x_\beta : \alpha < \beta\}$. The simple roots are $x_1 - x_2, x_2 - x_3, \dots, x_{n-1} - x_n$. It follows that the associated exterior differential system Σ defining the horizontal distribution is given by $\{\Omega_\beta^\alpha = 0 : 1 \leq \alpha < \beta \leq n, \beta \neq \alpha + 1\}$ on M.

We consider a *horizontal* holomorphic immersion $f : S \to M$ from a connected Riemann surface S.

Let e be a moving frame along f and ϕ, a type $(1,0)$ local coframe on S. Now define local complex valued functions $(Z_\beta^\alpha), 1 \leq \alpha < \beta \leq n$, as before by $e^* \Omega_\beta^\alpha = Z_\beta^\alpha \phi$.

Notation. $e^* \Omega_\beta^\alpha = \omega_\beta^\alpha$.

Using the horizontality of f in conjuction with the equations in Σ we get

$$\omega_\beta^\alpha = 0, \quad 1 \leq \alpha < \beta \leq n, \quad \beta \neq \alpha + 1,$$
$$\omega_\beta^{\beta-1} = Z_\beta^{\beta-1} \phi, \quad 2 \leq \beta \leq n. \tag{†}$$

Let $\tilde{e} = e \cdot t, t = (e^{it_1}, \dots, e^{it_n}) : U \cap \tilde{U} \to T$, be another moving frame along f and define $\tilde{Z}$ using $\tilde{e}$. Then we get the transformation rules

$$\tilde{Z}_\beta^{\beta-1} = e^{i(t_\beta - t_{\beta-1})} Z_\beta^{\beta-1}, \quad \beta \leq 2.$$

Put $r_\beta = Z_\beta^{\beta-1} \bar{Z}_\beta^{\beta-1}, \beta \geq 2$. (r_β) are globally defined invariants on S.

Definition. A point $x \in S$ is called a regular point of f if none of invariants (r_β) vanishes at x. Otherwise x is called a singular point of f.

Assigning S the metric induced by f, we assume that f is an isometric immersion. Without loss of generality we take ϕ to be unitary. It follows that $r_2 + r_3 + \ldots + r_n = 1$.

Proposition. $Z_\beta^{\beta-1} : U$ $(=$ the domain of $e) \to \mathbf{C}, \beta \geq 2$, is of analytic type.

Proof. Exterior differentiate both sides of the second set of equations in (†) (using the Maurer-Cartan structure equations) and obtain

$$[(\omega_{\beta-1}^{\beta-1} - \omega_\beta^\beta)Z_\beta^{\beta-1} + dZ_\beta^{\beta-1} + i\omega Z_\beta^{\beta-1}] \wedge \phi = 0, \quad \beta \geq 2,$$

where ω is the Levi-Civita connection form relative to ϕ. Rewriting,

$$dZ_\beta^{\beta-1} \equiv -Z_\beta^{\beta-1}(i\omega - \omega_\beta^\beta + \omega_{\beta-1}^{\beta-1})(\mathrm{mod}\ \phi).$$

Chern's theorem does the rest. q.e.d.

Lemma. Near a regular point of f we can choose a moving frame e such that relative to e, $Z_\beta^{\beta-1} > 0$ for every $\beta \geq 2$.

Proof. Looking at the transformation rules we see that in fact using only $S(U(1)^n)$-action $(t_1 + t_2 + \ldots + t_n = 1)$ we can accomplish the desired task. q.e.d.

Choose a moving frame e along f as in the lemma and write a_β instead of $Z_\beta^{\beta-1}$.

Remember that (r_β) are all globally defined nonnegative smooth functions on S and that $\Sigma r_\beta (2 \leq \beta \leq n) = 1$. Moreover, the above proposition tells us that each r_β has only isolated zeros (unless of course it is identically zero.) It follows that the positive square root of each $r_\beta(= a_\beta)$ is a function on S smooth away from the zeros and continuous at the zeros.

Definition. Let $f : S \to U(n)/T$ be a holomorphic horizontal isometric immersion from a connected Riemannian surface S. Then we define its Gauss-Frenet map $\Phi_f : S \to S^{n-2} < \mathbf{R}^{n-1}$ by $\Phi_f = (a_\beta)$.

Note that the Gauss-Frenet map is continuous everywhere and is smooth away from the singular points of f.

Convention. f as in the above definition will be called simply a horizontal holomorphic curve hereafter.

We now take care of the degenerate case thus simplifying the subsequent exposition.

Definition. A horizontal holomorphic curve $f : S \to M = U(n)/T$ is said to be generate if $r_\beta(f) \equiv 0$ for some $\beta \geq 2$.

Proposition. Let f be a degenerate horizontal holomorphic curve in $U(n)/T$. Then the image $f(S)$, upon applying a fixed unitary transformation (congruence), lies in $U(n-1)/T \cap U(n-1)$, where $U(n-1)$ is included in $U(n)$ by $g \mapsto \begin{pmatrix} g & 0 \\ 0 & 1 \end{pmatrix}$.

Proof. This is an elementary application of the Frobenius theorem on completely integrable systems. We take $n = 3$ and show that $f(S)$, upon a congruence, lies in $U(2)/T \cap U(2) \cong CP^1$. The rest of the proof, being quite analogous, is omitted. Looking at the first set of equations in (†) we see that $f(S)$ is congruent to an integral manifold of the exterior system on $G = U(3)$ given by $\{\Omega_2^1 = 0, \Omega_3^1 = 0\}(r_2(f) = 0)$, or $\{\Omega_3^2 = 0, \Omega_3^1 = 0\}(r_3(f) = 0)$. These systems are both completely integrable with maximal integral submanifolds cosets of $U(2) < U(3)$. Now a standard argument shows that $f(S)$ is congruent to an open submanifold of $U(2)/T \cap U(2)$. q.e.d.

For the rest of this section we exclude the degenerate curves from our consideration.

Let $f : S \to U(n)/T$ be a (nondegenerate) horizontal holomorphic curve. There is the Gauss-Frenet map of f given by $\Phi_f = (a_\beta) : S \to S^{n-2}$. A singular point x of f is where $a_\beta(x) = 0$ for some $\beta \geq 2$. (Remember that β runs from 2 to n.) These points are isolated since

(r_β) are all of analytic type. Near a regular point (†) becomes

$$\omega_\beta^\alpha = 0, \quad 1 \le \alpha < \beta \le n, \ \beta \ne \alpha + 1,$$
$$\omega_\beta^{\beta-1} = a_\beta \phi, \quad 2 \le \beta \le n. \tag{††}$$

Remark. Consider the exterior system on $S \times G$ given by $\{\Omega_\beta^\alpha = 0, \Omega_\gamma^{\gamma-1} = Z_\gamma \phi, 1 \le \alpha < \beta \le n, \beta \ne \alpha + 1, 2 \le \gamma \le n, (Z_\gamma)$ local complex valued functions on $S\}$ with independence condition $\phi \wedge \bar\phi \ne 0$. (Here we have the obvious abuse of notation, $\phi = \pi_1^* \phi$, where $\pi_1 : S \times G \to S$ etc.) It is then a straightforward matter to show that horizontal holomorphic curves in $U(n)/T$ correspond to the admissible integrals of the above system. It is in this way the equations (††) characterize f.

As we saw exterior differentiation of the second set of equations in (††) leads to

$$[da_\beta + i a_\beta \omega + a_\beta(\omega_{\beta-1}^{\beta-1} - \omega_\beta^\beta)] \wedge \phi = 0.$$

Observe that the first set of equations in (††), upon exterior differentiation yield nothing.

Rewriting the above equations,

$$[d\log a_\beta + i(\omega + i(\omega_\beta^\beta - \omega_{\beta-1}^{\beta-1}))] \wedge \phi = 0.$$

Noting that $\omega + i(\omega_\beta^\beta - \omega_{\beta-1}^{\beta-1})$ are real we get

$$*d\log a_\beta = \omega + i(\omega_\beta^\beta - \omega_{\beta-1}^{\beta-1}), \quad \beta \ge 2, \tag{*}$$

where $*$ is the Hodge operator of S, ds^2.

Let K denote the Gaussian curvature of (S, ds^2) so that $d\omega = \frac{i}{2} K \phi \wedge \bar\phi$. We also have $d * d\log a = \frac{i}{2}\Delta\log a \phi \wedge \bar\phi$, where Δ is the Laplace-Beltrami operator of (S, ds^2).

Using the Maurer-Cartan structure equations of $U(n)$ coupled with (††) we obtain $d\omega_\beta^\beta = (r_{\beta+1} - r_\beta)\phi \wedge \bar\phi$, $\beta \ge 2, r_1 = r_{n+1} = 0$.

Exterior differentiation of both sides of the equations in (*) now gives

$$\Delta \log a_\beta = K + 2r_{\beta-1} - 4r_\beta + 2r_{\beta+1},$$

or equivalently

$$\Delta \log r_\beta = 2K + 4r_{\beta-1} - 8r_\beta + 4r_{\beta+1}, \qquad (**)$$

where $2 \leq \beta \leq n, r_1 = r_{n+1} = 0.$

Consider the following coefficient matrix relative to $(r_{n-1}, \ldots, r_2,$ const) of the right hand size of $(**)$. (Here we use the relation $\Sigma r_\beta = 1$.)

$$CM = 4 \begin{bmatrix} & & & 1 & -2 & 0 \\ & & 1 & -2 & 1 & 0 \\ & & \vdots & & & \\ -3 & 0 & -1 & \ldots & -1 & -1 & 1 \\ 3 & 2 & 2 & \ldots & 2 & 2 & -2 \end{bmatrix}$$

The rank of CM is n. It follows that there exists exactly one affine relation amongst $\Delta \log r_2, \Delta \log r_3, \ldots, \Delta \log r_n$ and K. We write down this relation for $n = 3, 4$. (In general this relation is obtained by inverting CM.) Note that the case $n = 2$ is trivial since then $f(S)$ lies in $U(2)/U(1) \times U(1) \cong CP^1$.

$$n = 3 : \quad \Delta \log r_2 r_3 = 4(K - 1), \qquad (R3)$$

$$n = 4 : \quad \Delta \log r_2^9 r_3^2 r_4^3 = 4(7K - 4). \qquad (R4)$$

The following quantization theorem is now possible.

Proposition. Let $f : S \to U(n)/T$ be a horizontal holomorphic curve from a compact surface S. Then i) for $n = 3, K \geq 1$ implies that $K \equiv 1$, and ii) for $n = 4, K \geq \frac{4}{7}$ implies that $K \equiv \frac{4}{7}$.

Proof. We will prove i), ii) being similar. $K \geq 1$ says that $\Delta \log r_2 r_3 \geq 0$. So $\log r_2 r_3$ is a subharmonic function on S with singularities at the singular points of f where it goes to $-\infty$. Since S is assumed to be compact $\log r_2 r_3$ must attain a maximum in S. Hence it is constant by the maximum principle for subharmonic functions. It follows that $K \equiv 1$. q.e.d.

In the following we assume that S is compact and integrate (**).
As in §17 we have

$$\text{area}(S) = \frac{i}{2} \int_S \phi \wedge \bar{\phi},$$

$$\chi_S = \frac{i}{4\pi} \int_S K \phi \wedge \bar{\phi},$$

$$\#(r_\beta) = -\frac{i}{4\pi} \int_S \Delta \log r_\beta \phi \wedge \bar{\phi}.$$

$\#(r_\beta)$ denotes the number of zeros of r_β each counted with multiplicity. This number is, of course, finite since r_β is of analytic type and f is nondegenerate.

Integration of (**) over S yields

$$\#(r_\beta) + 2\chi_S = \frac{1}{\pi i} \int_S (r_{\beta-1} - 2r_\beta + r_{\beta+1}) \phi \wedge \bar{\phi},$$

where $2 \le \beta \le n, r_1 = r_{n+1} = 0, r_2 + \ldots + r_n = 1$.

Recalling the earlier discussion we see that the above integral formulae imply exactly one linear relation amongst $\#(r_\beta), \chi_S$ and the area of the immersion. For example, we have

Theorem. Let f be a compact nondegenerate horizontal holomorphic curve in $U(n)/T$. Then i) for $n = 3, 4\chi_S + \#(r_2) + \#(r_3) = \frac{2}{\pi} \text{area}(S)$, and ii) for $n = 4, 28\chi_S + 9\#(r_2) + 2\#(r_3) + 3\#(r_4) = \frac{8}{\pi} \text{area}(S)$.

§19. Holomorphic Curves in CP^m

CP^m denotes the complex projective space of complex dimension m which we think of as complex lines in $\mathbb{C}^{m+1}$. As a complex homogeneous space $CP^m = U(m+1)/U(1) \times U(m)$.

There is the projection $U(m+1)/U(1)^{m+1} = F_{1,2,\ldots,m+1}(\mathbb{C}^{m+1}) \to CP^m$ induced by the inclusion $U(1)^{m+1} \hookrightarrow U(1) \times U(m)$.

Let $G = U(m+1), H = U(1) \times U(m)$. We have the orthogonal decomposition (with respect to the negative of the Cartan-Killing form, say) of the Lie algebra $g, g = h \oplus m, m = h^\perp$.

$\Omega = (\Omega_\beta^\alpha), 1 \le \alpha, \beta \le m+1$, is the $u(m+1)$-valued Maurer-Cartan form of G. It decomposes into $\Omega = \Omega_h + \Omega_m$.

Restricting the invariant complex structure J (in §17) on $F_{1,2,\ldots,m+1}$ $(\mathbf{C}^{m+1})$ to CP^m we obtain an integrable invariant almost complex structure, denoted by J'. (Of course CP^m being hermitian symmetric the only other possibility is the conjugate structure.)

Then the "$(1,0)$-component" forms of the m-component of the Maurer-Cartan form are given by

$$\Theta^{1\beta} = \Omega_\beta^1, \quad 2 \le \beta \le m+1.$$

Let $g : S \to CP^m$ be a holomorphic isometric immersion from a connected Riemann surface with a fixed metric in its conformal class. From now on g will be called simply a holomorphic curve in CP^m.

On S we have the metric $ds_S^2 = \phi\bar{\phi}$, where ϕ is a local unitary type $(1,0)$ coframe. On CP^m we use a $U(m+1)$-invariant metric given by the pullback of $\Sigma\Omega_\beta^1\bar{\Omega}_\beta^1)(2 \le \beta \le m+1)$. (This is the normalized Fubini-Study metric.)

If E is a local moving frame along g then the holomorphy of g is reflected by the fact that $E^*\Omega_\beta^1(2 \le \beta \le m+1)$ are all type $(1,0)$. $\mathfrak{m} \cong \mathbf{C}^m$ via $\begin{bmatrix} 0 & -^t\bar{v} \\ v & 0 \end{bmatrix} \mapsto v \in \mathbf{C}^m$. The isotropy representation $i : U(1) \times U(m) \to GL(m)$ is given by

$$i(e^{it}, A) : v \mapsto e^{-it}Av, \quad (e^{it}, A) \in U(1) \times U(m).$$

Let E and $\tilde{E} = E \cdot t$ (t, a local $U(1) \times U(m)$-valued function on S) be two moving frames along g. We then have the formula $\tilde{E}^*\Omega_m = i(t^{-1})E^*\Omega_m$. From this it easily follows that there exists a moving frame E (about any point in S) with

$$E^*\Omega_2^1 = \phi, \quad E^*\Omega_\beta^1 = 0, \quad \beta > 2. \tag{1}$$

Notation. $E^*\Omega_\beta^\alpha = \omega_\beta^\alpha$.

The totality of frames achieving (1) is given by $E \cdot t$, where t is a local $U(1)^2 \times U(m-1)$-valued function on S. (Here we think of ϕ as an expression defined only up to modulus.)

Upon exterior differentiation (1) yields

$$\omega_\beta^2 = Z_\beta^2\phi, \quad \beta > 2, \quad i\omega = \omega_2^2 - \omega_1^1, \tag{2}$$

where (Z_β^2) are local complex valued functions on S and ϕ is the Levi-Civita connection form of (S, ds^2) relative to ϕ.

Computation with the restricted (to $U(1)^2 \times U(m-1)$) isotropy action reveals that we can now choose a moving frame so that in addition to (2) we have

$$\omega_3^2 = Z_3^2 \phi, \quad \omega_\beta^2 = 0, \quad \beta > 3.$$

Such frames are determined up to the structure group $U(1)^3 \times U(m-2)$.

Observations.

i) To avoid redundant considerations we may restrict ourselves to looking at "linearly full" holomorphic curves, i.e., those curves $g : S \to CP^m$ which do not lie in any CP^k for $k < m$.

ii) Though Z_3^2 is only locally defined, the expression $Z_3^2 \bar{Z}_3^2$ is independent of the choice of moving frames (up to the structure group $U(1)^3 \times U(m-2)$), hence defines a global invariant $r_3 = Z_3^2 \bar{Z}_3^2 : S \to \mathbf{R}$.

iii) For linearly full curves r_3 cannot be identically zero. Then a consideration similar to those in §18 shows that r_3 is of analytic type.

Exterior differentiation of the last equation in (2) leads to

$$K = 4 - 2r_3,$$

where K is the Gaussian curvature of S, ds^2.

Recursively proceeding we obtain a $U(1)^{m+1}$-reduction, $F \to S$, of the $U(1) \times U(m)$-principal bundle $g^{-1}G \to S$. Local sections of this bundle $F \to S$ are called the Frenet frames along g. If E is a Frenet frame along g then relative to E we have

$$\omega_{\lambda_1}^1 = \omega_{\lambda_2}^2 = \ldots = \omega_{\lambda_{m-1}}^{m-1} = 0, \quad \lambda_i > i+1 \qquad (\dagger)$$

and

$$\omega_2^1 = \phi, \omega_3^2 = Z_3^2 \phi, \ldots, \omega_{m+1}^m = Z_{m+1}^m \phi. \qquad (\dagger\dagger)$$

Put $r_\beta = Z_\beta^{\beta-1} \bar{Z}_\beta^{\beta-1}, \beta \geq 3$. Then (r_β) are globally defined nonnegative functions on S and none of them is identically zero for a linearly full curve and they are all of analytic type. We note that near a point in S where $r_3 r_4 \ldots r_{m+1} \neq 0$ one can choose a Frenet frame relative to which $(Z_\beta^{\beta-1})$ simultaneously become positive real valued.

Exterior differentiation of the relations in (††) yields

$$i\omega = \omega_2^2 - \omega_1^1, \quad i\omega_3 = \omega_3^3 - \omega_2^2, \ldots, i\omega_{m+1} = \omega_{m+1}^{m+1} - \omega_m^m,$$

where ω_β is the Levi-Civita connection form of the β-th osculating metric $ds_\beta^2 = r_\beta \phi \bar{\phi}$.

Rewriting the above equations (excluding the first one) we get

$$[d \log r_\beta + 2i(\omega - \omega_\beta)] \wedge \phi = 0, \quad \beta \geq 3.$$

This in turn implies that

$$2(\omega - \omega_\beta) = *d \log r_\beta, \quad \beta \geq 3.$$

Using the equality $d * d \log r = \frac{i}{2} \Delta \log r \phi \wedge \bar{\phi}$ as before, exterior differentiation of the above equations leads to

$$\Delta \log r_\beta = 2K + 4r_{\beta-1} - 8r_\beta + 4r_{\beta+1},$$

where $3 \leq \beta \leq m + 1$, $r_2 = 1, r_{m+2} = 0$.

The above relations correspond to (**) of §18. On a compact surface they can be integrated to yield integral formulae similar to those in §18. We leave this as an excercise for the reader. Our computation should be compared to that in [Gr] pp. 796–797.

$$U(m+1)$$

$$S \xrightarrow{\quad g \quad} CP^m \xleftarrow{\quad \pi_2 \quad} U(m+1)/T$$

Observe that the bundle of Frenet frames along g defines a holomorphic map $S \to U(m+1)/T$: If $E, \tilde{E}$ are any two Frenet frames along g then $\pi_1 \circ E = \pi_1 \circ \tilde{E}$ and consequently they define a global map $S \to U(m+1)/T$. It is also easy to see that the map hence defined is holomorphic with respect to the complex structure J with $\Theta^{\alpha\beta} = \Omega^\alpha_\beta, 1 \le \alpha < \beta \le (m+1)$, on $U(m+1)/T$. (J is an integrable extension of the complex structure J' on CP^m.) Moreover the equations in (†) imply that this lifting into $U(m+1)/T$ is actually horizontal. (To see this note that the equations in (†) correspond to the equations in (†) of §18 with $n = m+1$.)

The lifting into $U(m+1)/T$ of g is also an immersion and induces on S the metric $(1 + r_3 + \ldots + r_{m+1})\phi\bar{\phi}$. So the lifting increases the area.

Therefore the Frenet frame construction for holomorphic curves in CP^m amounts to horizontally and holomorphically (increasing the area by the factor of $1 + r_3 + \ldots + r_{m+1}$) lifting these curves to the flag manifold $U(m+1)/T$.

We now prove the converse of the above statement.

Theorem. Equip $U(m+1)/T$ with the invariant complex structure J as in the preceding discussion. Also equip CP^m with the restriction complex structure J' coming from J. Let $f : S \to U(m+1)/T$ be a nondegenerate horizontal holomorphic immersion from a Riemann surface S. Then the map $\pi_2 \circ F : S \to U(m+1)/T \to CP^m$ is a linearly full holomorphic branched immersion with isolated branch points (at the set specified in the proof.)

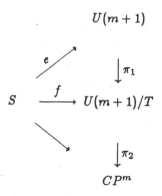

Proof. We will prove the theorem for $m = 2$, the general case being similar. We already saw that f holomorphic implies that $\pi_2 \circ f$ is holomorphic. Let e be a (local) moving frame along f. f holomorphic and horizontal means (§18) that $e^* \Omega_2^1 = X\phi, e^* \Omega_3^2 = Y\phi, e^* \Omega_3^1 = 0$ for some complex valued local functions X, Y and a local type $(1,0)$ coframe ϕ on S. Also f is an immersion iff $|X|^2 + |Y|^2$ is never zero. (Observe that the induced metric on S by f is $(|X|^2 + |Y|^2)\phi\bar\phi$.) Since f is nondegenerate the zeros of $|X|^2$ (and $|Y|^2$, which are globally defined) are at most isolated. But e is also a moving frame along the map $\pi_2 \circ f$ and it follows that $\pi_2 \circ f$ has isolated branch points at the zero set of $|X|^2$. (The induced metric on S by $\pi_2 \circ f$ is $|X|^2\phi\bar\phi$.) Finally $|Y|^2$ not identically zero implies that $\pi_2 \circ f$ is linearly full. q.e.d.

The above theorem together with the preceding discussion say roughly that horizontal curves in $U(m+1)/T$ correspond to holomorphic curves in CP^m via the projections and liftings associated with the Frenet frames.

§20. Horizontal Curves in $Sp(n)/T$

Let $G = Sp(n)$ and also let $T = U(1)^n$. (We think of the quaternions **H** as $\{z_1 + z_2 j : z_i \in \mathbf{C}\}$. This induces the inclusion $U(1)^n \hookrightarrow Sp(n)$.) Recall that $G = \{X \in GL(n; \mathbf{H}) : {}^t\bar{X}X = I\}$, where $\bar{\ }$ denotes the quaternionic conjugation. $M = G/T$ is the $Sp(n)$-flag manifold of real dimension $2n^2$.

There is the fibration with the standard fibre $(Sp(1)/U(1))^n = CP^1 \times \ldots \times CP^1, G/T \to Sp(n)/Sp(1)^n = F_{1,2,\ldots,n}(\mathbf{H}^n)$ (the full quaternionic flags in $\mathbf{H}^n$).

We have the usual decomposition $\mathfrak{g} = \mathfrak{t} \oplus \mathfrak{m}$. $\mathfrak{t}$, the Lie algebra of T, consists of purely imaginary (real multiples of $\sqrt{-1}$) n by n diagonal matrices. Let $E_{\alpha\beta} = $ the $n \times n$ matrix with $+1$ at (α, β)-entry, -1 at (β, α)-entry and zeros elsewhere. Here $1 \le \alpha < \beta \le n$. Also let $F_{\alpha\beta} = $ the $n \times n$ matrix with $+1$ at (α, β) and (β, α)-entries and zeros elsewhere. Here $1 \le \alpha \le \beta \le n$.

Put $V_{\alpha\beta} = RE_{\alpha\beta} \oplus iRF_{\alpha\beta}, 1 \le \alpha < \beta \le n$. Also put $V'_{\alpha\beta} = jRF_{\alpha\beta} \oplus kRF_{\alpha\beta}, 1 \le \alpha \le \beta \le n$. Then $\mathfrak{m} = \oplus\Sigma V_{\alpha\beta}(1 \le \alpha < \beta \le n) \oplus \Sigma V'_{\alpha\beta}(1 \le \alpha \le \beta \le n)$. Let $S(n; \mathbf{C})j$ denote the set of all complex symmetric matrices multiplied on the right by j (j is a unit quaternion). Then we can write

$$m = \left(u(n) \backslash t \right) \oplus S(n; \mathbf{C})j.$$

Let $t = \left(e^{it_\alpha} \right) \in T = U(1)^n$. The adjoint representation of T on m is given by

$$\mathrm{Ad}_t : v_{\alpha\beta} = xE_{\alpha\beta} + iyF_{\alpha\beta}(\varepsilon V_{\alpha\beta}) \mapsto e^{i(t_\alpha - t_\beta)}v_{\alpha\beta},$$
$$v'_{\alpha\beta} = xF_{\alpha\beta}j + iyF_{\alpha\beta}j(\varepsilon V'_{\alpha\beta}) \mapsto e^{i(t_\alpha + t_\beta)}v'_{\alpha\beta}.$$

Now $t \cong \mathbf{R}^n$ via $\mathrm{diag}(ix_1, \dots, ix_n) \mapsto (x_1, \dots, x_n)$. Then the roots $(< t^*)$ are $\{\pm 2x_\alpha, \pm(x_\alpha + x_\beta), \pm(x_\alpha - x_\beta), 1 \le \alpha < \beta \le n\}$. For positive roots we take $\Delta_+ = \{2x_\alpha, x_\alpha + x_\beta, x_\alpha - x_\beta, 1 \le \alpha < \beta \le n\}$. Then the simple roots are $\Delta_s = \{x_1 - x_2, x_2 - x_3, \dots, x_{n-1} - x_n, 2x_n\}$.

$\Omega = (\Omega_\beta^\alpha)$ denotes the $sp(n)$-valued Maurer-Cartan form of G. We note that $sp(n)$ consists of $n \times n$ **H**-valued matrices with ${}^tX = -X$ and hence $\Omega_\beta^\alpha = -\bar{\Omega}_\alpha^\beta$.

Let $\Gamma_\beta^\alpha, \Sigma_\beta^\alpha$ be the complex valued 1-forms with

$$\Omega_\beta^\alpha = \Gamma_\beta^\alpha + \Sigma_\beta^\alpha j.$$

Then $\Omega_\beta^\alpha = -\bar{\Omega}_\beta^\alpha$ is equivalent to $\Gamma_\beta^\alpha = -\bar{\Gamma}_\alpha^\beta, \Sigma_\beta^\alpha = \Sigma_\alpha^\beta$. That is to say, Γ is $u(n)$-valued and Σ is $S(n; \mathbf{C})$-valued.

The Maurer-Cartan structure equations $d\Omega = -\Omega \wedge \Omega$ become

$$d\Gamma_\beta^\alpha = -\Gamma_\gamma^\alpha \wedge \Gamma_\beta^\gamma + \Sigma_\gamma^\alpha \wedge \bar{\Sigma}_\beta^\gamma,$$
$$d\Sigma_\beta^\alpha = -\Gamma_\gamma^\alpha \wedge \Sigma_\beta^\gamma - \Sigma_\gamma^\alpha \wedge \bar{\Gamma}_\beta^\gamma.$$

Let J be the invariant (integrable) complex structure on G/T arising from the choice Δ_+, of positive roots. The corresponding "(1,0)-component" forms of Ω_m are $\Theta^{\alpha\beta} = \Gamma_\beta^\alpha, 1 \le \alpha < \beta \le n$, and $\Theta'^{\alpha\beta} = \Sigma_\beta^\alpha, 1 \le \alpha \le \beta \le n$.

For an invariant hermitian metric on G/T we take $-1/(2n + 2)$ times the Cartan-Killing form restricted to $m : ds^2 = -\mathrm{tr}(\Omega_m \cdot \Omega_m) = \Sigma \Gamma_\beta^\alpha \bar{\Gamma}_\beta^\alpha \ (1 \le \alpha < \beta \le n) + \Sigma \Sigma_\beta^\alpha \bar{\Sigma}_\beta^\alpha (1 \le \alpha \le \beta \le n)$.

Let S be a connected Riemann surface. We consider a holomorphic immersion $f : S \to Sp(n)/T$. If e is a local moving frame along f then we define $(Z_\beta^\alpha), (Z_\beta'^\alpha)$, local complex valued functions on S by

$$e^* \Gamma_\beta^\alpha = Z_\beta^\alpha \phi, \quad 1 \leq \alpha < \beta \leq n,$$
$$e^* \Sigma_\beta^\alpha = Z_\beta'^\alpha \phi, \quad 1 \leq \alpha \leq \beta \leq n, \tag{1}$$

where ϕ is a local type $(1,0)$ coframe on S.

If $\tilde{e} = e \cdot t$ $(t = (e^{it_\alpha})$ is a local T-valued function on $S)$ is another moving frame along f then define tilded·quantities by $\tilde{e}^* \Gamma_\beta^\alpha = \tilde{Z}_\beta^\alpha \phi, \tilde{e}^* \Sigma_\beta^\alpha = \tilde{Z}_\beta'^\alpha \phi$.

The adjoint representation of $T = U(1)^n$ quickly gives the transformation rules

$$\tilde{Z}_\beta^\alpha = e^{i(t_\beta - t_\alpha)} Z_\beta^\alpha, \quad 1 \leq \alpha < \beta \leq n,$$
$$\tilde{Z}_\beta'^\alpha = e^{-i(t_\alpha + t_\beta)} Z_\beta'^\alpha, \quad 1 \leq \alpha \leq \beta \leq n.$$

$r_\beta^\alpha = Z_\beta^\alpha \bar{Z}_\beta^\alpha$ and $r_\beta'^\alpha = Z_\beta'^\alpha \bar{Z}_\beta'^\alpha$ are globally defined invariants on S.

Definition. A point $x \in S$ is called a regular point of f if none of the invariants $(r_\beta^\alpha), (r_\beta'^\alpha)$ vanishes at x. Otherwise x is called a singular point of f.

Notation. $e^* \Gamma = \gamma, e^* \Sigma = \sigma$, etc.

Assume now that f is horizontal. Consulting the simple roots Δ_s above and the associated exterior system Σ we see that

$$\sigma_B^A = 0, \quad 1 \leq A, B \leq n, \quad (A, B) \neq (n, n),$$
$$\gamma_D^C = 0, \quad 1 \leq C < D \leq n, \quad D \neq C + 1. \tag{2}$$

Putting (1) and (2) together we get

$$\gamma_{i+1}^i = Z_i \phi, 1 \leq i \leq n - 1, (Z_i) \text{local complex valued functions on } S,$$
$$\sigma_n^n = Z_n \phi, Z_n, \text{a local complex valued function on } S, \tag{$\dagger$}$$
$$\sigma_B^A = 0, \gamma_D^C = 0, A, B, C, D \text{ as in the above.}$$

Notation. $r_i = Z_i \bar{Z}_i : S \to \mathbf{R} \geq 0, 1 \leq i \leq n$.

Assign S the metric induced by f and take ϕ to be unitary. It follows that $\Sigma \, r_i (1 \leq i \leq n) = 1$.

Proposition. $Z_i : U$ (= the domain of e) $\to \mathbf{C}$ is of analytic type for every i.

Proof. Exterior differentiate both sides of the equations in the first two lines of (†) (using the Maurer-Cartan structure equations) and obtain

$$[dZ_1 + iZ_1\omega + Z_1(\gamma_1^1 - \gamma_2^2)] \wedge \phi = 0,$$

$$[dZ_{n-1} + iZ_{n-1}\omega + Z_{n-1}(\gamma_{n-1}^{n-1} - \gamma_n^n)] \wedge \phi = 0, \qquad (3)$$

$$[dZ_n + iZ_n\omega + 2Z_n\gamma_n^n] \wedge \phi = 0,$$

where ω is the Levi-Civita connection form relative to ϕ. The rest follows. q.e.d.

Lemma. Near a regular point of f we can choose a moving frame e such that relative to e, $Z_i > 0$ for every i.

Proof. The transformation rules say $\tilde{Z}_i = e^{\sqrt{-1}(t_{i+1} - t_i)} Z_i$ for $1 \leq i \leq n - 1$ and $\tilde{Z}_n = e^{-2\sqrt{-1}t_n} Z_n$. So choose t_n to make $\tilde{Z}_n$ positive and then choose t_{n-1} to make $\tilde{Z}_{n-1}$ positive and so on. q.e.d.

Choose a moving frame e along f as in the lemma and write a_i instead of Z_i.

Definition. Let $f : S \to Sp(n)/T$ be a holomorphic horizontal isometric immersion from a connected Riemannian surface S. Then we define its Gauss-Frenet map $\Phi_f : S \to S^{n-1} < \mathbf{R}^n$ by $\Phi_f = (a_i)$.

Note that the Gauss-Frenet map is continuous everywhere and is smooth away from the singular points of f as in §18.

We will call f as in the above definition simply a horizontal holomorphic curve from now on.

We now take care of the degenerate case.

Definition. A horizontal holomorphic curve $f : S \to Sp(n)/T$ is said to be degenerate if $r_i(f) \equiv 0$ for some i.

Proposition. Let $f : S \to Sp(n)/T$ be a degenerate horizontal holomorphic curve. Then i) if $r_n \equiv 0$ then $f(S)$ is congruent to an open submanifold of $U(n)/T$, and ii) if $r_i \equiv 0 (i \neq n)$ then $f(S)$ is congruent to an open submanifold of $Sp(k)/T'$, where T' is a maximal torus of $Sp(k), k < n$.

Proof. This proof is completely similar to that in §18 and we give here only a sketch. Take $n = 3, r_1 \equiv 0$ for example. Then $f(S)$ (upon a congruence) $< Sp(2)/T'$: Just consider the exterior system on G given by $\{\Gamma_2^1 = \Gamma_3^1 = 0, \Sigma_1^1 = \Sigma_2^1 = \Sigma_3^1 = \Sigma_2^2 = \Sigma_3^2 = 0\}$. This system is completely integrable and its analytic subgroup is easily seen to be a subgroup of $H = U(1) \times Sp(2) < Sp(3)$. So $f(S)$ is congruent to an open submanifold of $H/T = \{e\} \times Sp(2)/\{e\} \times U(1)^2 = Sp(2)/T'$. q.e.d.

Remark. Note that a horizontal holomorphic curve in $Sp(n)/T$ is distinguished from a horizontal holomorphic curve in $U(n)/T$ by a single invariant, namely r_n. This somewhat surprising phenomenon (considering that $\dim Sp(n)/T \gg \dim U(n)/T$) is due to the fact that the rank of the horizontal distribution of $Sp(n)/T$ is only one greater than that of $U(n)/T$.

As we saw the exterior differentiation of (†) led to (3). (Observe that the equations in the last line of (†), upon exterior differentiation yield nothing.) Rewrite the equations in (3) as follows:

$$[d\log a_1 + i(\omega + i(\gamma_2^2 - \gamma_1^1)] \wedge \phi = 0,$$

$$\vdots$$

$$[d\log a_{n-1} + i(\omega + i(\gamma_n^n - \gamma_{n-1}^{n-1})] \wedge \phi = 0,$$

$$[d\log a_n + i(\omega - 2i\gamma_n^n)] \wedge \phi = 0.$$

Nothing that $\omega + i(\gamma_{i+1}^{i+1} - \gamma_i^i), \omega - 2i\gamma_n^n$ are all real, we get

$$* \, d\log a_1 = \omega + i(\gamma_2^2 - \gamma_1^1),$$

$$\cdot$$
$$\cdot$$
$$\cdot$$
$$\tag{*}$$

$$* \, d\log a_{n-1} = \omega + i(\gamma_n^n - \gamma_{n-1}^{n-1}),$$
$$* \, d\log a_n = \omega - 2i\gamma_n^n,$$

where $*$ is the Hodge operator of S, ds^2.

Using the Maurer-Cartan structure equations of $Sp(n)$ together with
(†) we obtain $d\gamma_1^1 = r_1\phi\wedge\bar\phi$, $d\gamma_2^2 = (r_2-r_1)\phi\wedge\bar\phi, \ldots, d\gamma_n^n = (r_n-r_{n-1})$
$\phi\wedge\bar\phi$. Exterior differentiation of both sides of the equations in (*) now
gives

$$\Delta\log a_1 = K + 2(r_2 - 2r_1),$$
$$\Delta\log a_2 = K + 2(r_3 - 2r_2 + r_1),$$

$$\cdot$$
$$\cdot$$
$$\cdot$$

$$\Delta\log a_{n-1} = K + 2(r_n - 2r_{n-1} + r_{n-2}),$$
$$\Delta\log a_n = K - 4(r_n - r_{n-1}),$$

or more compactly,

$$\Delta\log r_i = 2K + 4r_{i-1} - 8r_i + 4r_{i+1}, \tag{**}$$

where $r_0 = 0, r_{n+1} = r_{n-1}, K$ denotes the Gaussian curvature of
(S, ds^2), and Δ denotes the Laplace-Beltrami operator of (S, ds^2).

As in §18 there exists exactly one affine relation amongst $\Delta\log r_1$,
$\ldots, \Delta\log r_n$ and K. We write down this relation for $n = 2, 3$.

$$n = 2: \quad \Delta\log r_1^4 r_2^3 = 2(7K - 4), \tag{R2}$$

$$n = 3: \quad \Delta\log r_1^3 r_2^5 r_3^3 = 2(11K - 2). \tag{R3}$$

Corollary. Let $f : S \to Sp(n)/T$ be a horizontal holomorphic isometric immersion from a compact surface S. Then i) for $n = 2, K \geq \frac{4}{7}$
implies that $K \equiv \frac{4}{7}$, and ii) for $n = 3, K \geq \frac{2}{11}$ implies that $K \equiv \frac{2}{11}$.

Proof. Use the maximum principle for subharmonic functions. See
§18. q.e.d.

We record the integral formulae for compact S. (The derivation is similar to the one in §18.)

$$\#(r_i) + 2\chi_S = \frac{1}{\pi i} \int_S (r_{i-1} - 2r_i + r_{i+1})\phi \wedge \bar{\phi},$$

where $1 \leq i \leq n$.

Integrating (R2) and (R3) directly we get

$$n = 2: \quad 4\#(r_1) + 3\#(r_2) + 14\chi_S = \frac{4}{\pi}\text{area}(S),$$

$$n = 3: \quad 3\#(r_1) + 5\#(r_2) + 3\#(r_3) + 22\chi_S = \frac{2}{\pi}\text{area}(S).$$

Chapter V

ALGEBRAIC CURVES

In the preceding chapter, we have considered holomorphic immersions of Riemann surfaces. As it turns out, much of what we did there can be generalized to include branched holomorphic maps of Riemann surfaces into G-flag manifolds. The key idea is to allow "analytic type" singularities in the induced metric on a Riemann surface. In this we closely follow Griffiths. See [G-H] Chapter 2 §4.

A motivation for this chapter comes from algebraic geometry. Let $f : M \to CP^n$ be a nonconstant holomorphic map from a compact Riemann surface M. Then by Chow's theorem $f(M)$ is an algebraic curve. Conversely, if we let $C \subset CP^n$ be an algebraic curve then there exists a compact Riemann surface M and a holomorphic map $f : M \to CP^n$ such that $f(M) = C$. In fact M and f can be so chosen that outside the singular points of C, f becomes a biholomorphism. ($f : M \to CP^n$ is then called a nonsingular model of C. See [N] p. 275.) Now holomorphic curves in CP^n correspond to horizontal curves in $U(n+1)/T$ via the Frenet lifting. Thus the study of horizontal curves in $U(n+1)/T$ really is the study of projective algebraic curves.

In the present chapter we consider horizontal holomorphic maps from a compact Riemann surface into G/T. In particular we explicitly derive the Plücker formulae for horizontal curves in $Sp(n+1)/T$ (§25). It is apparent from our method that the Plücker formulae for horizontal curves in G/T depend essentially only on the Maurer-Cartan structure equations of G. The reader should be able to derive, for example, the formulae for $G = SO(m; \mathbf{R})$. We mention that the resulting orthogonal Plücker formulae "contain" as a special case the analogous formulae obtained by Chern [Ch1] for minimal two-spheres in a positive space form.

Exceptional groups present more challenge since to do computation with them one needs to look at their linear representations. It would be very interesting to explicitly write down the exceptional Plücker formulae.

§21. Singular Metrics and the Gauss-Bonnet-Chern Theorem

A brief review of the one-dimensional hermitian geometry follows. A hermitian metric on a Riemann surface M can be written locally as

$$ds^2 = h(z)dzd\bar{z} .$$

where z is a local holomorphic coordinate in M and h, a positive smooth local function on M. Putting $\varphi = \sqrt{h}\, dz$ we can also write

$$ds^2 = \varphi\bar{\varphi}$$

φ is then a local unitary coframe on M.

Write $\varphi = \varphi^1 + i\varphi^2$ with φ^1, φ^2 real 1-forms. We then have

$$d\varphi^1 = -\theta \wedge \varphi^2 ,$$
$$d\varphi^2 = \theta \wedge \varphi^1 ,$$

where θ is the Levi-Civita connection form relative to (φ^i) of (M, ds^2). The above pair of equations can be rewritten as the single complex equation,

$$d\varphi = i\theta \wedge \varphi = -\theta_{\mathbf{C}} \wedge \varphi .$$

$-i\theta = \theta_{\mathbf{C}}$ is the complex connection form of (M, ds^2) relative to φ.

The Gaussian curvature K of (M, ds^2) satisfies

$$d\theta_{\mathbf{C}} = \frac{K}{2}\, \varphi \wedge \bar{\varphi} .$$

The Kähler form of (M, ds^2) is by definition $-\frac{1}{2}\text{Im}\, ds^2$ which is equal to $\frac{i}{2}\, \varphi \wedge \bar{\varphi}$. It follows that

$$d\theta_{\mathbf{C}} = (-iK) \cdot \text{the Kähler form.} \tag{†}$$

$d\theta_{\mathbf{C}}$ is called the curvature form of (M, ds^2).

Suppose M to be compact of genus g. Then the Gauss-Bonnet-Chern theorem can be stated as

$$\frac{i}{2\pi} \int_M d\theta_{\mathbf{C}} = 2 - 2g \ .$$

Note that by (†) we get $2 - 2g = \frac{i}{4\pi} \int_M K\varphi \wedge \bar{\varphi}$ as before.

Definition: Let M be a Riemann surface. Then a singular metric on M is given by $ds^2 = \varphi\,\bar{\varphi}$, where φ is a type $(1,0)$ C^∞ form of analytic type, i.e., φ is locally written as the product of an analytic type function (see §17 for the definition) and a nowhere vanishing type $(1,0)$ form.

Let ds^2 be a singular metric on M. Then we can write $ds^2 = \varphi\,\bar{\varphi} = h(z)dz\,d\bar{z}$, where $h(z) \geq 0$ and z is a local holomorphic coordinate in M. Moreover, we have $h(z) = |z|^{2n}h_0(z)$, where $h_0(z)$ is never zero and n, a nonnegative integer. n is the order of φ at $z = 0$ and we write $\mathrm{ord}_o\,\varphi = n$. The singular divisor of ds^2, denoted by D_φ, is defined to be the zero divisor of φ. So $D_\varphi = \sum_{p \in M} \mathrm{ord}_p(\varphi)p$. It is easy to see that D_φ does not depend on the particular choice of φ but only on the singular metric given. The degree of D_φ is by definition $\sum_{p \in M} \mathrm{ord}_p(\varphi)$. This number is finite since φ is of analytic type and is the total number of zeros of φ counted with multiplicity.

The following equations are as before away from the zeros of φ which are isolated:

$$d\varphi = -\theta_{\mathbf{C}} \wedge \varphi \ ,$$
$$d\theta_{\mathbf{C}} = \frac{K}{2}\,\varphi \wedge \bar{\varphi} = (-iK) \cdot \text{the Kähler form}$$

The Gauss-Bonnet-Chern theorem generalizes as follows:

Theorem. Let M be a compact Riemann surface of genus g equipped with a singular metric $\varphi\,\bar{\varphi}$. Then

$$\frac{i}{2\pi} \int_M d\theta_{\mathbf{C}} = 2 - 2g + \deg D_\varphi \ .$$

A proof of this can be found in [G-H] p. 268. We just mention that $d\theta_{\mathbf{C}}$ is a multiple of $\Delta \log h \, dz \wedge d\bar{z}$ and the above theorem then follows from the usual Gauss-Bonnet-Chern theorem combined with the argument principle. The reader should try a direct proof.

§22. Projective Curves and Their Associated Curves

Throughout this section M denotes a compact Riemann surface of genus g. Let $f: M \to CP^n$ be a linearly full holomorphic map. Remember that this means $f(M)$ does not lie in any lower dimensional projective subspace. In what follows we discuss the associated curves of f following Griffiths (cf [G-H] pp. 263–286).

In a neighborhood of M, f can be holomorphically lifted to $\mathbf{C}^{n+1}\backslash \{0\}$. Let $v(z) = {}^t(v^0(z), \dots, v^n(z))$ be a such lifting. So v^i's are holomorphic and $[v(z)] = f(z)$. Put $e_0 = v, e_1 = v'$. Suppose we have another lifting $\tilde{v}$ of f into $\mathbf{C}^{n+1}\backslash\{0\}$. Then we must have $\tilde{v} = {}^t(\lambda v^0, \dots, \lambda v^n)$ for some C^∞ $\mathbf{C}^*$-valued local function λ on M. Put $\tilde{e}_0 = \tilde{v}, \tilde{e}_1 = \tilde{v}'$. Then $\tilde{e}_1 = {}^t(\lambda' v^i + \lambda v^{i'}) = {}^t(\lambda' v^i) + \lambda e_1$. It follows that

$$\tilde{e}_0 \wedge \tilde{e}_1 = \lambda^2 e_0 \wedge e_1 \tag{$\dagger$}$$

Note that the zero set of $e_0 \wedge e_1$ coincides with that of $\tilde{e}_0 \wedge \tilde{e}_1$. The following proposition is straightforward.

Proposition. f linearly full implies that the zero set of $e_0 \wedge e_1$ is isolated.

Whenever not zero $e_0 \wedge e_1$ defines a complex two-plane, which we denote by $[e_0 \wedge e_1]$ ($= \mathbf{C}$-span$\{e_0, e_1\}$), in $\mathbf{C}^{n+1}$. Moreover, ($\dagger$) says that $[e_0 \wedge e_1]$ only depends on f not on the choice of a homogeneous lifting. Put $f_1 = [e_0 \wedge e_1]: M\backslash\Sigma \to \mathbf{C}\,G_{n+1,2}$, where Σ is the zero set of $e_0 \wedge e_1$.

The Plücker embedding $i : \mathbf{C}\,G_{n+1,k+1} \to CP^N$, $N = \binom{n+1}{k+1} - 1$, is defined as follows: let $x = [E_0 \wedge \dots \wedge E_k] \in \mathbf{C}\,G_{n+1,k+1}$, $E_i \in \mathbf{C}^{n+1}\backslash\{0\}$. Fix a basis $\{\varepsilon_0, \dots, \varepsilon_n\}$ of $\mathbf{C}^{n+1}$ and write $E_0 \wedge \dots \wedge E_k = \sum_\alpha P_{\alpha_0 \dots \alpha_k} \varepsilon_{\alpha_0} \wedge \dots \wedge \varepsilon_{\alpha_k}$. Then $i: x \mapsto [P_{\alpha_0 \dots \alpha_k}]$.

Using the Plücker embedding we have $f_1 : M\backslash\Sigma \to CP^N$, $N = \binom{n+1}{2} - 1$. The following lemma is technically important.

Lemma. Let $F : U(\text{open in } M) \to \mathbf{C}^{N+1}$ be a holomorphic map with isolated zeros at $\Sigma \subset U$ so that $[F] = f : U \backslash \Sigma \to CP^N$ is a holomorphic map. Then there exists a unique holomorphic map $\hat{f} : U \to CP^N$ with $\hat{f}|_{U \backslash \Sigma} = F$.

Proof. $F : z \mapsto {}^t(v^0(z), \ldots, v^N(z))$, v^i's holomorphic. Suppose $z_0 \in \Sigma$, i.e., $v^i(z_0) = 0$ for every i. Let $k = \min_i \text{ord}_{z_0} v^i$. Consider the map $\hat{F} : z \mapsto {}^t(z^k v^0(z), \ldots, z^k v^N(z))$. Then $\hat{F}$ is holomorphic and $\hat{F}(z_0) \neq 0$. Put $\hat{f} = [\hat{F}]$. q.e.d.

Applying the lemma to $f_1|_{M \backslash \Sigma}$ we see that there exists a unique holomorphic extension of f_1 to all of M which we call once again f_1.

Definition. $f_1 : M \to \mathbf{C}\,G_{n+1,2} = CP^{n*}(\mathbf{C}\,G_{n+1,2}$ is identified with the set of projective lines in CP^n) is called the dual curve of f or the first associated curve of f.

Maintaining the above notation let $e_0 = {}^t(v^i)$, $e_1 = {}^t(v^{i'})$, $e_2 = {}^t(v^{i''})$. There are also the tilded quantities: $\tilde{e}_0 = {}^t(\tilde{v}^i) = {}^t(\lambda v^i)$, $\tilde{e}_1 = {}^t(\lambda' v^i + \lambda v^{i'})$, $\tilde{e}_2 = {}^t(\lambda'' v^i + 2\lambda' v^{i'} + \lambda v^{i''})$. It follows at once that $[\tilde{e}_0 \wedge \tilde{e}_1 \wedge \tilde{e}_2] = [e_0 \wedge e_1 \wedge e_2]$ whenever $e_0 \wedge e_1 \wedge e_2 \neq 0$ so that $[\ ]$ is defined. The linear fullness of f once again guarantees that the zero set of $e_0 \wedge e_1 \wedge e_2$ (which coincides with the zero set of $\tilde{e}_0 \wedge \tilde{e}_1 \wedge \tilde{e}_2$) is isolated. By the lemma the map $[e_0 \wedge e_1 \wedge e_2]$ is holomorphically extended to all of M.

Definition. $f_2 = [e_0 \wedge e_1 \wedge e_2]: M \to \mathbf{C}\,G_{n+1,3}$ is called the second associated curve of f.

Recursively proceeding we obtain

$$f_k = [e_0 \wedge \ldots \wedge e_k] : \quad M \to \mathbf{C}\,G_{n+1,k+1} \subset CP^N \ ,$$

where $N = \binom{n+1}{k+1} - 1$, $0 \leq k \leq n - 1$, and $f_0 = f = [e_0]: M \to CP^n$, a linearly full holomorphic map.

Definition. Let $f; M \to CP^n$ be a linearly full holomorphic map. Then the k-th associated degree of f, denoted by d_k, is the degree of $f_k(M)$ considered as an algebraic curve in $CP^N, N = \binom{n+1}{k+1} - 1$.

f as in the above. Using the inhomogeneous coordinates write $f(z) = [1, f^1(z), \ldots, f^n(z)]$. The ramification index at z_0 of f, denoted by $\#(z_0)$, is defined to be $\min_i \operatorname{ord}_{z_0}(f^{i'})$. We also put $\# = \sum_{z \in M} \#(z)$. $\#$ is the total ramification index of f. Similarly define $\#_k(z_0), \#_k$ for $f_k : M \to CP^N$. Note that $\#(z_0) \neq 0$ iff z_0 is a branch point of f.

The proof of the following observation is left as an exercise.

Observation. Suppose $f : M \to CP^n$ is a linearly full holomorphic immersion. Then $\#_k = \frac{1}{2}\#(r_k)$ with $\#(r_k)$ as in §§18–19.

§23. Horizontal Curves and the Frenet Frames

Let $f = f_0 : M \to CP^n = CG_{n+1,1}$ be a linearly full holomorphic map from a compact Riemann surface M. In the previous section, we have constructed the associated curves of f which are holomorphic maps $f_k : M \to CG_{n+1,k+1}$, $1 \leq k \leq n-1$. Now let $e = (e_0, \ldots, e_0)$ be a unitary frame along f. That is, $e : U$ (open in M) $\to U(n+1)$ with $[e_0] = f$.

Definition. A unitary frame e along f is called a Frenet frame along f if

$$[e_0 \wedge \ldots \wedge e_k] = f_k \quad \text{for every } k . \tag{†}$$

We remark that the notion of a Frenet frame along f in the above coincides with the one given in §19 when f is an immersion.

Let $\tilde{e}$ be another Frenet frame along f. (†) says that the complex k-plane spanned by the first k vectors of $\tilde{e}$ at $z \in \tilde{U} \subset M$ has to be equal to $f_k(z)$ which is also the complex k-plane spanned by the first k vectors of e at z. This holds for every k and since $e, \tilde{e}$ are $U(n+1)$-valued we get $\tilde{e} = e \cdot t$ where $t : U \cap \tilde{U} \to T = U(1)^{n+1}$, a smooth map. Hence there arises a well-defined map $\Phi_f : M \to U(n+1)/T$.

Generalizing the theorem in §19 we obtain

Theorem. Let $f : M \to CP^n$ be a linearly full holomorphic map. Then the map $\Phi_f : M \to U(n+1)/T$ is a nondegenerate horizontal holomorphic map. Conversely, let $F : M \to U(n+1)/T$ be a nondegenerate horizontal holomorphic map. Then $[E_0] : M \to CP^n$ is a linearly full holomorphic map where $E = (E_0, \ldots, E_n)$ is any unitary frame along F.

Proof. Let $f : M \to CP^n$ be a linearly full holomorphic map and also let e be a Frenet frame along f so that (†) holds. Put $e^*\Omega = \omega$, where Ω denotes the Maurer-Cartan form of $U(n+1)$. (†) says that each e_k is a C-linear combination of $e_0, e'_0, \ldots, e_0^{(k)}$. Combining this fact with the equations $de_i = \omega_i^j \otimes e_j$ gives

$$\omega_j^i = 0, \qquad 0 \le i < j < n, \quad j \ne i+1. \qquad (a)$$

We also have as we saw in §19

$$\omega_{k+1}^k, \qquad 0 \le k \le n-1, \quad \text{all of type } (1,0) \qquad (b)$$

due to the holomorphy of f. Comparing (a) and (b) against (††) of §18 we see that Φ_f is a horizontal holomorphic map where we use the standard complex structure on $U(n+1)/T$ explicitly introduced in §18. The rest of the proof is left to the reader. q.e.d.

Remark. Let $F : M \to U(n+1)/T$ be a nondegenerate horizontal holomorphic map and also let $E = (E_0, \ldots, E_n)$ be any unitary frame along F. Then there are the maps $[E_i] : M \to CP^n$, $1 \le i \le n$. By Bryant's theorem, these maps are all harmonic.

§24. Plücker Formulae for Projective Curves

Let $f : M \to CP^n$ be a linearly full holomorphic map. We then have

$f_k : M \to C\,G_{n+1,k+1}$, $1 \le k \le n-1$, associated curves,

$e : U \subset M \to U(n+1)$, a Frenet frame along f,

$\omega = e^*\Omega$, Ω is the Maurer-Cartan form of $U(n+1)$, and the structure equations

$$\begin{cases} \omega_j^i = 0 & 0 \le i < j \le n, \quad j \ne i+1, \\ \omega_{k+1}^k, & 0 \le k \le n-1, \quad \text{type } (1,0) \text{ forms}. \end{cases} \qquad (*)$$

Slightly modifying the proof of the first proposition in §18 one proves the following lemma.

Lemma. ω_{k+1}^k, $0 \le k \le n-1$, are all of analytic type.

Definition. The k-th induced singular metric (or the k-th osculating singular metric) is defined to be $ds_k^2 = \omega_{k+1}^k \bar\omega_{k+1}^k$, for $0 \le k \le n-1$. Its Kähler form is $\Lambda_k = \frac{i}{2}\omega_{k+1}^k \wedge \bar\omega_{k+1}^k$.

Recall that $f_k : M \to CP^N$, $N = \binom{n+1}{k+1} - 1$. Let ds_N^2 denote the normalized Fubini-Study metric on CP^N given in §19. The following proposition is purely computational.

Proposition. $ds_k^2 = f_k^* ds_N^2$.

It follows that $\Lambda_k = f_k^*$ (the Kähler form of (CP^N, ds_N^2)).

Notation. $\varphi_k = \omega_{k+1}^k$, so $ds_k^2 = \varphi_k \bar\varphi_k$.

We also let D_k denote the singular divisor of ds_k^2 so that $\deg D_k = \#_k$, the total ramification index of f_k.

We wish to compute the connection form and the curvature form of (M, ds_k^2) relative to φ_k. Using the Maurer-Cartan structure equations of $U(n+1)$ and consulting the relations in (*) we obtain

$$d\varphi_k = d\omega_{k+1}^k = -\left(\omega_k^k - \omega_{k+1}^{k+1}\right) \wedge \varphi_k .$$

Put $\theta_k = \omega_k^k - \omega_{k+1}^{k+1}$. θ_k is the (complex) connection form of (M, ds_k^2).

Upon exterior differentiation,

$$d\theta_k = d\left(\omega_k^k - \omega_{k+1}^{k+1}\right) = -\varphi_{k-1} \wedge \bar\varphi_{k-1} + 2\varphi_k \wedge \bar\varphi_k - \varphi_{k+1} \wedge \bar\varphi_{k+1} .$$

Rewriting the above equations,

$$d\theta_k = 2i\left(\Lambda_{k-1} - 2\Lambda_k + \Lambda_{k+1}\right) . \tag{†}$$

We also have

$$d\theta_k = -iK_k\Lambda_k ,$$

where K_k is the Gaussian curvature of (M, ds_k^2).

By the Gauss-Bonnet-Chern theorem (remember that $M \cong Tg$),

$$\frac{i}{2\pi}\int_M d\theta_k = 2 - 2g + \#_k .$$

Now $\int_M \Lambda_k$ is the area of (M, ds_k^2) and the Writinger theorem says

$$\frac{1}{\pi}\int_M \Lambda_k = d_k(= \text{the degree of } f_k) .$$

The Plücker formulae now follow:

$$2g - 2 - \#_k = d_{k-1} - 2d_k + d_{k+1} \, ,$$

where $0 \le k \le n - 1$, $d_{-1} = d_n = 0$, $d_0 = \deg f$.

§25. The Symplectic Plücker Formulae

We consider a holomorphic map $f : M \to Sp(n + 1)/T$, where M is a Riemann surface. It is understood that on G-flag manifolds we use the standard complex structures explicitly introduced in Chapter IV.

Definition. f is said to be nondegenerate if neither $f(M) \subset Sp(k)/T'$ $(k < n+1$ and T', a maximal torus of $Sp(k))$ nor $f(M) \subset U(n+1)/T$.

Remarks. i) We use the standard inclusion $U(n + 1) \subset Sp(n + 1)$ induced from the inclusion $\mathbf{C} \subset \mathbf{H} = \mathbf{C} \oplus \mathbf{C} j$ given by $z \mapsto z + 0j$. ii) As usual we do not distinguish two maps that are congruent, i.e., $f, g : M \to Sp(n + 1)/T$ with $f = A \circ g$, $A \in Sp(n + 1)$, fixed.

A local C^∞ lifting $e = (e_0, \dots, e_n)$ into $Sp(n+1)$ of f will be called a symplectic frame along f. If $\tilde{e} = (\tilde{e}_i)$ is any other symplectic frame along f then on their common domain we must have $\tilde{e} = e \cdot t$, where t is a $T = U(1)^{n+1}$-valued C^∞ local function on M. Thus there are well-defined maps $f_i = [e_i] : M \to S^{4n+3}/U(1)$, S^{4n+3} = the unit vectors in $\mathbf{H}^{n+1}$.

$\Omega = (\Omega_j^i)$, $0 \le i, j \le n$, denotes the $sp(n+1)$-valued Maurer-Cartan form of $Sp(n + 1)$. Γ, Σ are the complex 1-forms with $\Omega = \Gamma + \Sigma j$. Then Γ is $\mu(n + 1)$-valued and Σ, $S(n + 1; \mathbf{C})$-valued.

The Maurer-Cartan structure equations of $Sp(n+1)$ can be written as

$$d\Gamma = -\Gamma \wedge \Gamma + \Sigma \wedge \bar{\Sigma} \, ,$$
$$d\Sigma = -\Gamma \wedge \Sigma - \Sigma \wedge \bar{\Gamma} \, .$$

Notation. $e^*\Omega = \omega, e^*\Gamma = \gamma, e^*\Sigma = \sigma$.

The holomorphy of f implies that

$$\gamma_j^i(0 \le i < j \le n), \quad \sigma_j^i(0 \le i \le j \le n) \quad \text{are all of type } (1,0) \, . \quad (\dagger)$$

From now on we assume that f is horizontal. Then as in §20 we have

$$\begin{cases} \gamma^i_j = 0, & 0 \le i < j \le n, \quad j \ne i+1, \\ \sigma^i_j = 0, & (i,j) \ne (n,n). \end{cases} \tag{††}$$

So the only possibly nonzero type $(1, 0)$ forms among (†) are γ^k_{k+1}, $0 \le k \le n-1$, σ^n_n.

Notation. $\varphi_k = \gamma^k_{k+1}$, $0 \le k \le n-1$, $\varphi_n = \sigma^n_n$.

It is by now routine to check that each $\varphi_k (0 \le k \le n)$ is of analytic type. Note also that $\bar{\varphi}_k \bar{\varphi}_k = \varphi_k \bar{\varphi}_k$ $(0 \le k \le n)$ from the transformation rule derived in §20. Put $\Lambda_k = \frac{i}{2}\varphi_k \wedge \bar{\varphi}_k$. As in §20 we have

Proposition. Let $f : M \to Sp(n+1)/T$ be a horizontal holomorphic map. Then f is nondegenerate if and only if none of the Λ_k's $(0 \le k \le n)$ are identically zero.

Observe that $f(M) \subset U(n+1)/T$ (hence it represents a projective curve) if and only if $\Lambda_n \equiv 0$.

We assume for the rest of this section that $f : M \to Sp(n+1)/T$ is a nondegenerate horizontal holomorphic map. It follows that each $ds^2_k = \varphi_k \bar{\varphi}_k$ $(0 \le k \le n)$ is a singular metric on M and Λ_k, its Kähler form.

Using the Maurer-Cartan structure equations and (††) we obtain

$$\begin{cases} d\varphi_k = -(\gamma^k_k - \gamma^{k+1}_{k+1}) \wedge \varphi_k & (k \ne n), \\ d\varphi_n = -2\gamma^n_n \wedge \varphi_n \end{cases}$$

Put $\theta_k = \gamma^k_k - \gamma^{k+1}_{k+1}, \theta_n = 2\gamma^n_n$. They are the complex connection forms.

Using the Maurer-Cartan structure equations and (††) again we obtain $d\gamma_0 = \varphi_0 \wedge \bar{\varphi}_0$, $d\gamma^k_k = -\varphi_{k-1} \wedge \bar{\varphi}_{k-1} + \varphi_k \wedge \bar{\varphi}_k$ $(k \ne 0)$. This gives

$$\begin{cases} d\theta_0 = 2i(-2\Lambda_0 + \Lambda_1), \\ d\theta_k = 2i(\Lambda_{k-1} - 2\Lambda_k + \Lambda_{k+1}), \\ d\theta_n = 2i(2\Lambda_{n-1} - 2\Lambda_n). \end{cases}$$

Or, more compactly,

$$d\theta_k = 2i(\Lambda_{k-1} - 2\Lambda_k + \Lambda_{k+1}), \tag{*}$$

where $0 \le k \le n$, $\Lambda_{-1} = 0$, $\Lambda_{n+1} = \Lambda_{n-1}$.

Suppose M to be compact of genus g. The integration of (*) over M now yields

$$2g - 2 - \#_k = \frac{1}{\pi} \int_M \Lambda_{k-1} - 2\Lambda_k + \Lambda_{k+1} ,$$

where $\#_k$ is the degree of the singular divisor of ds_k^2.

Notation. $d_k = \frac{1}{\pi} \int_M \Lambda_k \left(= \frac{1}{\pi} \cdot (\text{the area of } M, ds_k^2) \right)$.

It follows that

$$2g - 2 - \#_k = d_{k-1} - 2d_k + d_{k+1} \qquad (**)$$

where $0 \le k \le n$, $d_{-1} = 0$, $d_{n+1} = d_{n-1}$.

We call (**) the symplectic Plücker formulae. Let C denote the coefficient matrix ($(n+1)$ by $(n+1)$) of the RHS of (**). So,

$$C = \begin{bmatrix} -2 & 1 & 0 & 0 & \cdots & 0 & 0 \\ 1 & -2 & 1 & 0 & \cdots & 0 & 0 \\ 0 & 1 & -2 & 1 & \cdots & 0 & 0 \\ & & & \cdot & & & \\ & & & \cdot & & & \\ & & & \cdot & & & \\ & & & & & 0 & 2 & -2 \end{bmatrix} .$$

Then (**) can be rewritten as

$$2g - 2 - \#_k = C_k^i d_i .$$

Lemma. (d_k) are uniquely determined by $(\#_k)$ and vice versa.

Proof. The rank of C is $n + 1$. q.e.d.

In the following we give an application of the sympletic Plücker formulae.

Theorem. Let $f : M \cong T_g \to Sp(n+1)/T$ be an unramified nondegenerate horizontal holomorphic map. Then $g = 0$, i.e., $M \cong CP^1$.

Proof. We will give the proof for $n = 2$, the general case being similar. f, unramified says $\#_0 = \#_1 = \#_2 = 0$. So the symplectic Plücker formulae become $2g - 2 = C_k^i d_i$, $0 \le k \le 2$. Inverting the matrix C we get $d_0 = -5g + 5$, $d_1 = -8g + 8$, $d_2 = -9g + 9$. The nondegeneracy of f implies that $d_k > 0$ for every k. So $g = 0$, $d_0 = 5, d_1 = 8, d_2 = 9$. q.e.d.

Appendix

EXTERIOR DIFFERENTIAL SYSTEMS

As much of our exposition is couched in, explicitly or implicitly, the language of exterior differential forms, we give here an outline of the theory of exterior differential systems. Our aim is to give a concise and fairly accessible account of the theory so that it can be quickly applied to problems.

The first three sections of our notes dealing with the Cartan-Kähler theory are based upon the book by Cartan [Ca3] and the notes by Bryant-Chern-Griffiths [B-C-G]. The results here are given without proofs. Section 4 deals with the notion of prolongation with emphasis on overdetermined Pfaffian systems in two independent variables. This section follows closely [Y2] §1.

Our account is elementary and does not cover some of the more sophisticated tools of the trade such as characteristic sheafs or systems of finite type. The reader may consult a recent monograph [G-J] and the references cited therein.

§1. Exterior Algebra

Let V be a real vector space of dimension n and V^*, its dual. A bilinear form $F : V \times V \to \mathbf{R}$ is called alternating if $F(u, v) = -F(v, u)$ for every $u, v \in V$. If $e_1, \ldots, e_n$ is a basis of V and $u = u^i e_i, v = v^i e_i$ then $F(u, v) = a_{ij} u^i j^j$ where $a_{ij} = F(e_i, e_j) = -a_{ji}$.

The two-fold exterior product $\Lambda^2(V)$ is defined to be $V \otimes V/I$, where I is the subspace spanned by $\{u \otimes v + v \otimes u : u, v \in V\}$. We write $u \wedge v \in \Lambda^2(V)$, i.e., $u \wedge v$ is the equivalence class represented by $u \otimes v$. Note that $\{e_i \wedge e_j : 1 \leq i < j \leq n\}$ is a basis of $\Lambda^2(V)$. So dim $\Lambda^2(V)$ $= \binom{n}{2} = \frac{1}{2} n(n-1)$.

There is a natural vector space isomorphism $\Lambda^2(V^*) \to \{$alternating bilinear forms on $V\}$ given by $\alpha \wedge \beta(u, v) = \alpha(u)\beta(v) - \alpha(v)\beta(u)$, $\alpha, \beta \in V^*$.

For any bilinear form F on V and any vector $e \in V$, define the interior product $i_e F$ to be the 1-form $(\in V^*)$ $F(e, \cdot)$. Thus $(i_e F)(v) = F(e, v)$, $v \in V$. It follows that i_e induces an antiderivation on $\Lambda^2(V^*)$, that is, $i_e(\alpha \wedge \beta) = \alpha(e)\beta - \beta(e)\alpha$.

The following two propositions are basic.

Proposition (Cartan's Lemma). Let $\theta^1, \dots, \theta^p$ be p linearly independent 1-forms on V, i.e., elements of V^*. If p 1-forms $\varphi^1, \dots, \varphi^p$ satisfy $\theta^1 \wedge \varphi^1 + \dots + \theta^P \wedge \varphi^P = 0$ then $\varphi^i = a^i_j \theta^j$ for every i with $a^i_j = a^j_i$.

Proposition (Canonical Form). For any $F \in \Lambda^2(V^*)$ there exist $2p$ $(0 \le p \le n/2)$ independent 1-forms $\theta^1, \dots, \theta^{2p-1}, \theta^{2p} \in V^*$ such that $F = \theta^1 \wedge \theta^2 + \dots + \theta^{2p-1} \wedge \theta^{2p}$.

Let $p \ge 2$ be an integer. Also let $\tau \in S_p = $ the symmetric group on p letters. Define a p-linear map $f_\tau : V \times \dots \times V \to V \otimes \dots \otimes V = \otimes^p(V)$ by $f_\tau(v_1, \dots, v_p) = v_{\tau(1)} \otimes \dots \otimes v_{\tau(p)}$. By the universal mapping property f_τ induces a linear map $g_\tau : \otimes^p(V) \to \otimes^p(V)$. Define $A_p : \otimes^p(V) \to \otimes^p(V)$ by $A_p = \frac{1}{p!} \sum_{\tau \in S_p} \mathrm{sgn}(\tau) g_\tau$. Put $I_p = \ker A_p$. Then the p-fold exterior product of V is defined to be $\Lambda^p(V) = \otimes^p(V)/I_p$. $\Lambda^p(V) = 0$ if $p > n$ and $\dim \Lambda^p(V) = \binom{n}{p}$ for $0 \le p \le n$. Set $\Lambda(V) = \oplus_{p=0}^n \Lambda^p(V)$. $\Lambda(V)$ is a graded algebra over $\mathbf{R}$ of dimension 2^n, and is called the exterior algebra over V. Note that if $x \in \Lambda^p(V), y \in \Lambda^q(V)$ then $y \wedge x = (-1)^{p+q} x \wedge y$.

$\Lambda^p(V^*)$ is naturally identified with the space of p-linear alternating forms on V. Elements of $\Lambda^P(V^*)$ are called p-forms on V and elements of $\Lambda^p(V)$ are called p-vectors in V.

A homogeneous ideal in $\Lambda(V)$ is an ideal I such that $I = \oplus_{p=1}^n I_p$, where $I_p = I \cap \Lambda^P(V)$. If Σ is a subset of $\Lambda(V)$ consisting of homogeneous elements then $I(\Sigma)$ denotes the homogeneous ideal generated by Σ.

For $v \in V$ and a p-form F on V, define $i_v F(v_1, \dots, v_{p-1})$ to be $F(v, v_1, \dots, v_{p-1})$. $i_v F$ is a $(p-1)$-form on V. We have, for $\theta^1, \dots, \theta^p$

$\in V^*, i_v(\theta^1 \wedge \ldots \wedge \theta^p) = \sum_{i=1}^p (-1)^{i+1} \theta^i(v) \theta^1 \wedge \ldots \wedge \theta^{i-1} \wedge \theta^{i+1} \wedge \ldots \wedge \theta^p.$

Let $\Phi \in \Lambda^p(V^*)$ and also let W be a q-dimensional linear subspace in V (q-plane for short). Then Φ restricted to W, $\Phi|_W$, is the p-form on W given by $(w_1, \ldots, w_p) \mapsto \Phi(w_1, \ldots, w_p), w_i \in W$. We say that W is a solution to the equation $\Phi = 0$ if $\Phi|_W = 0$. Note that if $q < p$ then $\Phi|_W = 0$ trivially.

Definition. A system of exterior equations on V is a finite set Σ of homogeneous forms on V. A q-plane W in $\dot{V}$ is a solution of Σ if $\Phi|_W = 0$ for every $\Phi \in \Sigma$. We also say that such W annihilates Σ.

Let $I(\Sigma)$ be the homogeneous ideal in $\Lambda(V^*)$ generated by Σ. Clearly any q-plane which annihilates Σ also annihilates $I(\Sigma)$. We say that two systems Σ_1, Σ_2 are algebraically equivalent if $I(\Sigma_1) = I(\Sigma_2)$. Algebraically equivalent systems admit the same solutions.

Let Σ be a system of exterior equations on V. The associated space of Σ is the subspace of V given by $A_\Sigma = \{v \in V : i_v\Phi \in I(\Sigma),$ for every $\Phi \in \Sigma\}$. Its annihilator $A_\Sigma^\perp \subset V^*$ is called the dual associated space of Σ. Observe that $A_\Sigma = A_{I(\Sigma)}$.

A p-form Φ is said to be decomposable if it is a monomial in $\Lambda^p(V^*)$. It is easy to verify that a p-form Φ is decomposable if and only if dim $A_\Phi = n - p, n = \dim V$.

The significance of A_Σ is given by the following theorem.

Theorem. Let Σ_1 be a system of exterior equations on V. Then there exists a system of exterior equations Σ_2 algebraically equivalent to Σ_1 such that Σ_2 consists of elements of $\Lambda(A_{\Sigma_1}^\perp)$.

So the theorem gives "minimal" sets of generators of homogeneous ideals in $\Lambda(V^*)$.

§2. Completely Integrable Systems and the Cauchy Characteristics

Let M be a smooth (real) manifold of dimension n. At a point $x \in M$ let M_x denote the tangent space to M at x and M_x^*, its dual space (cotangent space). We also let $x^1, \ldots, x^n$ be a local coordinate system in M.

An exterior differential form ω on M assigns to each $x \in M$ an exterior form $\omega_x \in \Lambda(M_x^*)$. Locally, $\omega = \frac{1}{k!} \sum_i A_{i_1 \ldots i_k} dx^{i_1} \wedge \ldots \wedge dx^{i_k},$

$1 \leq i_1, \ldots, i_k \leq n$, where the coefficients are smooth functions and are antisymmetric in any two indices.

Let $\mathscr{E}(M) = \otimes_{k=0}^{n} \mathscr{E}^k(M)$ denote the set of all exterior differential forms on M, where $\mathscr{E}^k(M)$ is the set of all k-forms on M. Then $\mathscr{E}(M)$ is a graded algebra over $\mathscr{E}^0(M) = $ the ring of smooth functions on M. There exists a unique mapping, called the exterior differentiation, $d : \mathscr{E}(M) \rightarrow \mathscr{E}(M)$ which sends k-forms to $(k+1)$-forms characterized by the following properties:

i) $d(\omega + \theta) = d\omega + d\theta$, $\quad \omega, \theta \in \mathscr{E}(M)$,

ii) if $\omega \in \mathscr{E}^k(M)$ then $d(\omega \wedge \theta) = d\omega \wedge \theta + (-1)^k \omega \wedge d\theta$,

iii) on $\mathscr{E}^0(M)$, d is the ordinary differential of functions, and

iv) for $f \in \mathscr{E}^0(M)$, $d \cdot df = d^2 f = 0$.

It follows that $d^2 = 0$. One of the key properties of d is that it commutes with the pullback map.

Definition. An exterior differential system is a finite set Σ of homogeneous exterior differential forms on M. A solution to Σ, i.e., to the system of equations $\{\omega = 0, \omega \in \Sigma\}$ is a submanifold $f : S \rightarrow M$ such that $f^*\omega = 0$ for every $\omega \in \Sigma$. Such submanifolds are called integrals or integral submanifolds of the system.

If $I(\Sigma)$ is the homogeneous ideal in $\mathscr{E}(M)$ generated by Σ then $f : S \rightarrow M$ is an integral of Σ if $f^*\omega = 0$ for every $\omega \in I(\Sigma)$.

For any subset $A \subset \mathscr{E}(M)$ put $dA = \{d\omega : \omega \in A\}$. An exterior differential system Σ is said to be closed if $d\Sigma \in I(\Sigma)$ (or equivalently if $dI(\Sigma) \subset I(\Sigma)$). The closure of Σ is defined to be $\bar{\Sigma} = \Sigma \cup d\Sigma$. Algebraically equivalent systems have algebraically equivalent closures.

Let $f : S \rightarrow M$ be an integral of the system Σ. Then $df^*\omega = f^*d\omega = 0$ for any $\omega \in \Sigma$. Thus one may replace the original system Σ by its closure $\bar{\Sigma}$ without changing the integrals.

A theoretically crucial class of exterior differential systems is that of Pfaffian systems. A Pfaffian system on M is given by

$$\Sigma = \left\{ \theta^\alpha = 0 : \theta^\alpha \in \mathscr{E}^1(M), \alpha = 1, \ldots, s \right\}, \tag{1}$$

where we assume that 1-forms (θ^α) are linearly independent. Geometrically, Σ can be thought of as a subbundle of the linear frame bundle.

If one assumes that Σ is closed then

$$d\theta^\alpha \equiv 0 \quad (\mathrm{mod}\,\Sigma), \quad \alpha = 1, \ldots, s \,. \tag{2}$$

This condition is called the Frobenius condition, and a closed Pfaffian system is called a completely integrable system.

Theorem (Frobenius). Suppose that the Pfaffian system (1) is completely integrable. Then in a sufficiently small neighborhood there is a local coordinate system $y^1, \ldots, y^n$ such that $I(\Sigma)$ is generated by $dy^1, \ldots, dy^s$.

It follows that M is partitioned into maximal integrals $((n - s)$-dimensional submanifolds locally given by $y^1 = c_1, \ldots, y^s = c_s, c_i$ constant) of the closed Pfaffian system Σ. In such a case we say that the system defines a foliation of which the maximal integrals are called the leaves.

We now let Σ be an arbitrary closed exterior differential system on M. If possible, we like to replace Σ by an algebraically equivalent system which is simpler.

Recall that at $x \in M$,

$$(A_\Sigma)_x = \{v \in M_x : i_v \Sigma_x \subset I(\Sigma_x)\} \,.$$

A vector field η such that $\eta_x \in (A_\Sigma)_x$ is called a Cauchy characteristic vector field of Σ.

Theorem. Let Σ be a closed differential system on M such that A_Σ has constant dimension $n - s$. Then there exists a neighborhood about any point of M in which there are coordinates $(y^i; y^a),\ 1 \le i \le n - s$, $n - s + 1 \le a \le n$ such that $I(\Sigma)$ has a set of generators that are forms in (y^a) only. In particular, in such a case A_Σ is a foliation whose leaves are called the Cauchy characteristics.

Assume that the foliation given by A_Σ (assuming A_Σ to be of constant dimension) is a fibration $\pi : M \to N$ so that the fibres are the Cauchy characteristics. (This is always true locally.) So, locally $\pi : (y^i; y^a) \to (y^a)$. It follows that we may replace the original system Σ on M by the obvious system $\tilde{\Sigma}$ on N. To recover the original integrals it suffices to obtain integrals in N and look at their inverse images in M.

§3. Cartan-Kähler Theory

At the outset we emphasize that the theory presented here is local in nature and all data, real analytic. This means, for example, we replace the manifold under consideration by a neighborhood in it whenever necessary without any explicit mention of it. Also maps and objects are in the real analytic category.

Let Σ be a closed exterior differential system on an n-dimensional manifold M. Put $\Sigma^k = \Sigma \cap \mathscr{E}^k(M)$. Let $Z = \{x \in M : f(x) = 0, f \in \Sigma^0\}$. Also let Z^* denote the set of smooth points of the real analytic variety Z. Thus if $r = n - \dim Z^*$ and $x \in Z^*$ then there exist exactly r independent 1-forms $df_1, \ldots, df_r$ at x with $f_1, \ldots, f_r \in \Sigma^0$.

An integral p-element of Σ is a pair (E^p, x) where $x \in Z$ and E^p is a p-plane in M_x such that E^p annihilates Σ_x. $V_p(\Sigma) \subset G_p(M)$ $(=$ the Grassmann bundle of p-planes over $M)$ denotes the set of all integral p-elements of Σ. $V_p(\Sigma)$ is an analytic variety in $G_p(M)$.

The zeroth order character of Σ is

$$s_0 = \max_{x \in Z^*} \left(\dim \operatorname{span} \Sigma_x^1 \right) .$$

$x \in Z^*$ is called a regular integral point if $\dim \operatorname{span} \Sigma_x^1 = s_0$. Note that $Z^* \backslash \{\text{regular integral points}\}$ is the intersection of Z^* and a proper subvariety of Z. In particular, $x \in Z^*$ is regular iff near x $\dim \operatorname{span} \Sigma^1$ is constant.

An integral 1-element (E^1, x) is said to be ordinary if x is regular. Let (E^1, x) be an ordinary integral 1-element. Put $E_1 = \operatorname{span} \{e_1\}, e_1 \in M_x$. Then the first order character of Σ, denoted by s_1, is defined by the equation

$$s_0 + s_1 = \max \left(\dim \operatorname{span} \{ \Sigma_x^1 \cup i_{e_1} \Sigma_x^2 \} \right) ,$$

where the maximum is taken over all ordinary integral 1-elements (E^1, x).

The system of equations represented by $\Sigma_x^1 \cup i_{e_1} \Sigma_x^2$ is called the polar system of (E^1, x). An ordinary integral 1-element is said to be regular if the dimension of its polar system is $s_0 + s_1$. It is not hard to show that a regular integral 1-element is a smooth point of the variety $V_1(\Sigma)$.

An integral 2-element is said to be ordinary if it contains at least one regular integral 1-element. Note that in order for an ordinary integral 2-element to exist we must have $n - (s_0 + s_1) > 1$. (The converse is false.) Let (E^2, x) be an ordinary integral 2-element and put $E^2 = \text{span} \{e_1, e_2\} = [e_1 \wedge e_2]$. Then the second order character of Σ, denoted by s_2, is defined by the equation

$$s_0 + s_1 + s_2 = \max \left(\dim \text{span} \{\Sigma_x^1 \cup i_{e_1} \Sigma_x^2 \cup i_{e_2} i_{e_1} \Sigma_x^3\} \right),$$

where the maximum is taken over all ordinary integral 2-elements (E^2, x).

In general, an integral p-element (E^p, x) is called ordinary if it contains at least one regular integral $(p-1)$-element (E^{p-1}, x). Thus an ordinary integral p-element (E^p, x) gives rise to the flag at x

$$E^p \supset E^{p-1} \supset \ldots \supset E^1$$

where except possibly E^p every other element is regular.

$s_0 + s_1 + \ldots + s_p$ is the dimension of the polar system of a regular integral p-element which is also the maximum dimension of polar systems of ordinary integral p-elements. An ordinary $(p+1)$-element exists only if $n - (s_0 + \ldots + s_p) > p$.

Let $O_p(\Sigma)$ denote the set of ordinary integral p-elements and also let $R_p(\Sigma)$ denote the set of regular integral p-elements. So $R_p(\Sigma) \subset O_p(\Sigma) \subset V_p(\Sigma) \subset G_p(M)$.

Fact. $R_p(\Sigma)$ is dense in $O_p(\Sigma)$ and both $R_p(\Sigma)$ and $O_p(\Sigma)$ are Zariski-open in $V_p(\Sigma)$.

The genus of Σ is the first integer h such that there does not exist an ordinary integral $(h+1)$-element.

Theorem (Cartan-Kähler). Let Σ be a closed exterior differential system with genus h. Also let $(E^p, x), p \leq h$, be an ordinary integral p-element and let $g : S^{p-1} \to M$ be a $(p-1)$-dimensional integral submanifold passing through x and tangent at x to a regular integral $(p-1)$-element $E^{p-1} \subset E^p$. Then there exists an infinity of p-dimensional integral submanifolds containing g and tangent at x to E^p. Roughly speaking, the totality of these p-dimensional integral submanifolds is parameterized by $(n-p) - (s_0 + \ldots + s_{p-1})$ arbitrary functions of p variables.

Though the Cartan-Kähler theorem is of central theoretical importance it is rarely applied in its present form. Many applications require the existence of "admissible" integral submanifolds satisfying a transversality condition which we discuss presently.

Definition. A closed exterior differential system Σ on M together with a decomposible p-form, hereafter denoted by $\omega \in \mathscr{E}^p(M)$, is called a closed exterior differential system with independence condition (or with specified independent variables). An admissible integral element of such a system is an integral p-element of Σ on which $\omega \neq 0$. An admissible integral manifold is a p-dimensional integral whose tangent spaces are admissible.

So a p-dimensional integral $f : S \to M$ will be admissible iff $f^* \omega \neq 0$. Also note that ω may be defined modulo the system Σ.

A closed differential system with independence condition is said to be in involution at $x \in M$ if there exists an ordinary admissible integral p-element (E^p, x). In such a case there exist locally admissible integral manifolds by virtue of the Cartan-Kähler theorem.

Any system of partial differential equations can be thought of as a closed exterior differential system with specified independent variables. Consider, say, a second order system of PDE's in the (z^a) as functions of the (x^i), where $1 \leq i \leq p, 1 \leq a \leq n - p$. Put $t_i^a = \partial z^a / \partial x^i$, $t_{ij}^a = \partial z^a / \partial x^i \partial x^j$. Then this system of PDE's consists of equations relating $x^i, z^a, t_i^a, t_{ij}^a$. Define the exterior differential system Σ on (x, z, t)-space as follows:

$\Sigma^0 =$ the set of functions in x, z, t defining the system of PDE's,
$\Sigma^1 = \{ dz^a - t_i^a dx^i, dt_i^a - t_{ij}^a dx^j \}$,
$\Sigma^2 = d\Sigma^1$.

We take $\omega = dx^1 \wedge \ldots \wedge dx^p$. Admissible integrals are p-dimensional integral submanifolds given as graphs of maps $z^a = z^a(x), t_i^a = t_i^a(x)$.

We let $V_p(\Sigma, \omega) \subset V_p(\Sigma)$ denote the variety of admissible integral elements. Also put $O_p(\Sigma, \omega) = V_p(\Sigma, \omega) \cap O_p(\Sigma)$.

Let $\omega^1, \ldots, \omega^p$ be p independent 1-forms such that $\omega = \omega^1 \wedge \ldots \wedge \omega^p$. We define the reduced polar system of an integral element (E^q, x) of (Σ, ω) to be its polar system modulo $(\omega^1, \ldots, \omega^p)$. This definition depends only on ω not on the choice (ω^i).

Let X be an irreducible component of $V_p(\Sigma, \omega)$, i.e., X is a maximal subvariety of $V_p(\Sigma, \omega)$ that cannot be written as a union of two varieties. Then the reduced characters $s'_0, s'_1, \ldots, s'_{p-1}$ of (Σ, ω) with respect to X are defined by letting $s'_0, s'_0 + s'_1, \ldots, s'_0 + s'_1 + \ldots + s'_{p-1}$ be the maximal dimensions of the reduced polar systems of $(E^1, x), (E^2, x), \ldots, (E^{p-1}, x)$ respectively, where each integral element (E^i, x) is contained in a member of X. The symbol σ_p is defined by the formula

$$\sigma_p = (n - p) - (s'_0 + \ldots + s'_{p-1}) .$$

The following theorem is called Cartan's involutivity criterion.

Theorem. Let X be an irreducible component of $V_p(\Sigma, \omega)$. Then

$$ps'_0 + (p - 1)s'_1 + \ldots + s'_{p-1} \leq n + (n - p)p - \dim X .$$

Moreover, equality holds if and only if X contains ordinary integral elements, i.e., (Σ, ω, X) is in involution.

Note that the above inequality is equivalent to the inequality

$$\dim X \leq n + s'_1 + 2s'_2 + \ldots + p\sigma_p .$$

§4. Prolongation

From now on we restrict our discussion to the case of Pfaffian systems. This will simplify exposition without greatly sacrificing generality as it can be easily shown that the first prolongation of any system is Pfaffian. Another salient point to make is that the notion of prolongation applies only to those systems with independence condition.

Let M be a smooth manifold of dimension n. Consider the following Pfaffian system on M given locally as

$$\Sigma = \{\theta^\alpha = 0: \ \theta^\alpha \in \mathcal{E}^1(M), \ \alpha = 1, \dots, s\} \,, \tag{1}$$

where we assume that $\theta^1 \wedge \dots \wedge \theta^s \neq 0$.

Globally Σ is given by a subbundle $I \subset T^*M$ of rank s. So (θ^α) locally span I. An independence condition then is given by a subbundle $J(\supset I)$ of rank $s + p$ in T^*M. So locally $\{(\theta^\alpha, \omega^i): 1 \leq \alpha \leq s, 1 \leq i \leq p\}$ span J. Put $\omega = \omega^1 \wedge \dots \wedge \omega^p$. We want an integral submanifold $f: S \to M$ with $f^*\omega \neq 0$.

Index convention. $1 \leq \alpha, \beta, \gamma, \dots, \leq s$, $1 \leq i, j, k, \dots \leq p$, $1 \leq a, b, c, \dots \leq n - (p + s)$.

Since we are interested in admissible integrals only we may as well assume that $\underset{i}{\wedge} \omega^i \wedge \underset{\alpha}{\wedge} \theta^\alpha \neq 0$. (Otherwise admissible integrals clearly do not exist.) Now choose (π^a) such that $(\omega^i, \pi^a, \theta^\alpha)$ form a coframe on M.

Differentiate (θ^α) and, suppressing the terms containing (θ^α), obtain the following.

$$d\omega^\alpha \equiv A^\alpha_{ai} \pi^a \wedge \omega^i + \frac{1}{2} B^\alpha_{ij} \omega \wedge \omega^j + \frac{1}{2} C^\alpha_{ab} \pi^a \wedge \pi^b \ (\mathrm{mod}\,\Sigma) \,, \tag{2}$$

where $A^\alpha_{ai}, B^\alpha_{ij} = -B^\alpha_{ji}, C^\alpha_{ab} = -C^\alpha_{ba}$ are functions on M.

We say that Σ is in normal form if $(C^\alpha_{ab}) = 0$. If Σ is in normal form then after the following change of coframe on M, Σ is still in normal form:

$$\tilde{\theta}^\alpha = T^\alpha_\beta \theta^\beta \,,$$
$$\tilde{\omega}^i = T^i_j \omega^j + T^i_\alpha \theta^\alpha \,,$$
$$\tilde{\pi}^a = T^a_b \pi^b + T'^a_\beta \theta^\beta + T^a_j \omega^j \,,$$

where $T^\alpha_\beta, T'^a_\beta, T^i_j, T^a_b$ are nonsingular transformations.

Let (e_i, e_a, e_α) be the frame field in M dual to $(\omega^i, \pi^a, \theta^\alpha)$. We will choose a local coordinate system in $G_p(M)$ about the p-plane field defined by $e_1 \wedge \dots \wedge e_p$, arising from the choice $(\omega^i, \pi^a, \theta^\alpha)$. A p-plane

E in the neighborhood defined by $\omega^1 \wedge \ldots \wedge \omega^p \neq 0$ is represented by a matrix of the form

$$\begin{bmatrix} (\delta_j^i) \\ (l_j^a) \\ (l_j^\alpha) \end{bmatrix} .$$

To put it another way, E is defined by a decomposible p-vector $v_1 \wedge \ldots \wedge v_p$, where $v_j = e_j + l_j^a e_a + l_j^\alpha e_\alpha$. We take (l_j^a, l_j^α) as the standard fibre coordinates in $G_p(M)$.

We now express the condition that (θ^α) and $(d\theta^\alpha)$ vanish on E in terms of the above standard coordinates in $G_p(M)$. On such a $E \in G_p(M)$ we have $\theta^\alpha = l_j^\alpha \omega^j = 0$, $\pi^a = l_j^a \omega^j$ and $d\theta^\alpha = 0$. Using (2) and substituting we obtain

$$F_i^\alpha = l_i^\alpha = 0 , \tag{3}$$

$$F_{ij}^\alpha = A_{aj}^\alpha l_i^a - A_{ai}^\alpha l_j^a + B_{ij}^\alpha + C_{ab}^\alpha (l_i^a l_j^b - l_i^b l_j^a) = 0 . \tag{4}$$

The analytic variety in $G_p(M)$ defined by (3) and (4) is precisely $V(\Sigma, \omega)$, the variety of admissible integral p-elements. If $V(\Sigma, \omega)$ is empty then we say the system is incompatible. Observe that if Σ is in normal form then the equations in (3), (4) define a system of affine equations and hence the variety $V(\Sigma, \omega)$ is irreducible.

In the case of a Pfaffian system in normal form Cartan's involutivity criterion reduces to linear algebra. We now give a simple test for involutivity in the case of two independent variables. So $p = 2$. The equations in (1), (2) become

$$\Sigma = \{\theta^\alpha = 0, \theta^\alpha \in \mathscr{E}^1(M), \alpha = 1, \ldots, s\} ,$$
$$d\theta^\alpha \equiv A_{ai}^\alpha \pi^a \wedge \omega^i + B^\alpha \omega^1 \wedge \omega^2 \pmod{\Sigma} .$$

Also (3) and (4) become

$$F_i^\alpha = l_i^\alpha = 0 , \tag{5}$$

$$F^\alpha = A_{a2}^\alpha l_1^a - A_{a1}^\alpha l_2^a + B^\alpha = 0 . \tag{6}$$

Observe that $s_0' = s$. Put $\# = $ the number of defining equations of $V(\Sigma, \omega)$ in the fibre of $G_2(M)$. So $\# = 2(n-2) - [\dim V(\Sigma, \omega) - n]$.

(Remember that $V(\Sigma, \omega)$ is irreducible.) Now Cartan's criterion says that (Σ, ω) is in involution if and only if

$$2s + s_1' = \# \ .$$

Put r = the number of independent equations in (6). Then $\# = 2s + r$ since (5) contains $2s$ equations. Recall that s_1' is the rank of $i_e d(\theta^\alpha) \pmod{\omega^1, \omega^2}$ where e is a general integral vector contained in a member of $V(\Sigma, \omega)$. Formally, s_1' is the rank of the matrix

$$(P_b^\alpha) = (\partial A_{ai}^\alpha \pi^a \wedge \omega^i / \partial \pi^b) \ ,$$

where $\partial(\pi^a \wedge \pi^b)/\partial \pi^c = \delta_c^a \pi^b - \delta_c^b \pi^a$. (P_b^α) is called the reduced polar matrix of the system (Σ, ω). Submmarizing, we have

Theorem. Let (Σ, ω) be a Pfaffian system with two independent variables in normal form. Then it is in involution if and only if the number of independent "quadratic" equations equals the rank of the reduced polar matrix.

Going back to the arbitrary p case we define the notion of prolongation.

Definition. Let Σ be as in (1). Then the (first) prolongation of Σ, denoted by $\Sigma_{(1)}$, is the exterior differential system on $G_p(M)$ given by

$$\Sigma_{(1)} = \Sigma_{(1)}^0 \cup \Sigma_{(1)}^1 \cup \Sigma_{(1)}^2 \ , \text{where}$$
$$\Sigma_{(1)}^0 = \{F_i^\alpha = 0 \,, F_{ij}^\alpha = 0\} \ ,$$
$$\Sigma_{(1)}^1 = \{dF_i^\alpha = 0 \,, dF_{ij}^\alpha = 0 \,, \theta^\alpha = 0 \,, \pi^a - l_j^a \omega^j = 0\} \ ,$$
$$\Sigma_{(1)}^2 = d\Sigma_{(1)}^1 \ .$$

In the above we wrote θ^α in place of $\pi^* \theta^\alpha, \pi : G_p(M) \to M$, etc.

Note that $V(\Sigma, \omega)$ is precisely the variety defined by $\Sigma_{(1)}^0$. It is often convenient to look at the Pfaffian system $\hat{\Sigma}_{(1)}$ on $V(\Sigma, \omega)$ obtained from $\Sigma_{(1)}$ by restriction. $\hat{\Sigma}_{(1)}$ on $V(\Sigma, \omega)$ is sometimes called the canonical system. Note that $\hat{\Sigma}_{(1)}$ is always in normal form.

The Cartan-Kuranishi prolongation theorem states, roughly, that given an exterior differential system with independence condition it takes a finite number of prolongations for it to be either involutive or incompatible. The idea is by increasing the number of unknowns one introduces a possibility of converting singular integrals of the system into ordinary integrals of the prolonged system. (One notes that upon a prolongation the number of newly created unknowns is in general greater than the number of newly created equations.)

The following theorem gives a special case of the Cartan-Kuranishi theorem.

Theorem. Overdetermined normal Pfaffian systems with two independent variables in s equations and $s - 1$ unknowns require at the most $s - 1$ successive prolongations to be either involutive or incompatible.

For the proof see [Y2] pp. 472–473.

REFERENCES

[A]   J. F. Adams, *Lectures on Lie Groups*, Benjamin, New York
         1969.

[A-B-S] M. F. Atiyah, R. Bott and A. Shapiro, Clifford Modules,
         Topology **3** *Supplement 1 (1964) 3–38.*

[B-C-G] R. Bryant, S. Chern and P. Griffiths, *Notes on exterior
         differential systems*, Proc. 1980 Peking Symposium on Dif-
         ferential equations, Peking University 1980.

[B-D]   T. Bröcker and T. Dieck, *Representations of Compact Lie
         Groups*, Springer-Verlag, New York 1985.

[B-H]   A. Borel and F. Hirzebruch, Characteristic classes and
         homogeneous spaces I, *Am. J. of Math.* **80** (1958) 458–
         536.

[B1]   R. L. Bryant, Submanifolds and special structures on the
         octanians, *J. of Diff. Geom.* **17** (1982) 185–232.

[B2]   R. L. Bryant, Conformal and minimal immersions of com-
         pact surfaces into the 4-sphere, *J. of Diff. Geom.* **17** (1982)
         455–473.

[B3]   R. L. Bryant, Lie groups and twistor spaces, *Duke Math. J.*
         52 (1985) 223–262.

[Bu]   F. E. Burstall, A twistor description of harmonic maps of
         a 2-sphere into a Grassmannian, *Math. Ann.* **274** (1986)
         61–74.

[Ca1]   É. Cartan, *Théorie des Groupes Finis et Continus et la
         Géométrie Différentielle Traitées par la Méthode du Repère
         Mobile*, Gauthier-Villars, Paris 1937.

[Ca2]   É. Cartan, *Oeuvres Comlètes, Partie I*, Gauthier-Villars,
         Paris 1952.

[Ca3]   É. Cartan, *Les Systèmes Differential Extérieurs et Leurs
         Applications Géomètriques*, Hermann, Paris 1945.

[Ch1]   S. S. Chern, On the minimal immersions of the two-sphere in a space of constant curvature, *Problems in Analysis*, Princeton 1970, 27–40.

[Ch2]   S. S. Chern, *Selected Papers*, Springer-Verlag, New York 1978.

[Ch3]   S. S. Chern, *Complex Manifolds without Potential Theory*, Second Edition, Springer-Verlag, New York 1979.

[C-W]   S. S. Chern and J. G. Wolfson, Minimal surfaces by moving frames, *Am. J. of Math.* **105** (1983) 59–83.

[E-W]   J. Eells and J. C. Wood, Harmonic maps from surfaces to complex projective spaces, *Advances in Math.* **49** (1983) 217–263.

[G-H]   P. Griffiths and J. Harris, *Principles of Algebraic Geometry*, Wiley-Interscience, New York 1978.

[G-J]   P. Griffiths and G. Jensen, *Differential Systems and Isometric Embeddings*, Princeton University Press 1987.

[Gl]   J. F. Glazebrook, The construction of a class of harmonic maps to quaternionic projective spaces, *J. of London Math. Soc.* **30** (1984) 151–159.

[Gr]   P. Griffiths, On Cartan's method of Lie groups and moving frames as applied to uniqueness and existence questions in differential geometry, *Duke Math. J.* **41** (1974) 775–814.

[He]   S. Helgason, *Differential Geometry, Lie Groups and Symmetric Spaces*, Academic Press, New York 1978.

[Hu]   J. E. Humphreys, *Introduction to Lie algebras and Representation Theory*, Springer-Verlag, New York 1970.

[J]   G. R. Jensen, *Higher Order Contact of Submanifolds of Homogeneous Spaces*, Lecture Notes in Math. 610, Springer-Verlag, New York 1977.

[J-R-Y] G. R. Jensen, M. Rigoli and K. Yang, Holomorphic curves in the complex quadric, *Bull. of Austral. Math. Soc.* **35** (1978) 125–147.

[N]   M. Namba, *Geometry of Projective Algebraic Curves*, Dekker, New York, 1984.

[S]   S. Salamon, Quaternionic Kähler manifolds, *Invent. Math.* **67** (1982) 143–171.

[Wan]   H. C. Wang, Closed manifolds with homogeneous complex structures, *Am. J. of Math.* **76** (1954) 1–32.

[War]   F. Warner, *Foundations of Differentiable Manifolds and Lie Groups*, Scott, Foresman and Company, Glenview 1971.

[Y1]   K. Yang, Frenet formulae for holomorphic curves in the two quadric, *Bull. of Austral. Math. Soc.* **33** (1986) 195–206.

[Y2]   K. Yang, Deformation of submanifolds of real projective space, *Pacific J. of Math.* **120** (1985) 469–492.

INDEX

Adjoint map 6
Admissible integral 98
Analytic type function 60, 63, 79, 81
Associated curves 82, 83

Canonical form 92
Cartan subalgebra 13
Cartan-Kähler theory 96, 98
Cartan-Killing form 8, 37
Cartan-Kuranishi theorem 102, 103
Cartan's involutivity criterion 99
Cartan's lemma 92
Cayley-Graves algebra 1
Characters 96
    reduced characters 99
Chow's theorem 79
Complementary roots 34
Compact Lie group 1
    biinvariant metrics on 10, 14, 28
    connected abelian group 6
    examples of 3, 4
    left invariant metrics on 10
    Lie algebra of 5
    one parameter subgroup 5
    semisimple 8
    simple 8
    weight of 22
Complex quadric 55
Complex submanifold 45

Degree 86, 89
Division algebra 1
Dynkin diagram 22, 24, 26

E_6 25, 26
E_7 25, 26

E_8 25, 26
Euler-Lagrange equations 57, 66, 77
Euler-Poincare characteristic 6, 61
Exponential map 6
Exterior differential system 44, 65
   Cauchy characteristics of 93
   closed 95
   completely integrable 93
   PDE 98
   Pfaffian 94, 102
Exterior equation 93

F_4 25, 26
Frenet frame 71, 84
Frobenius theorem 1, 95
Fubini-Study metric 38

G_2 25, 26
G_2-flag manifold 56
G-flag manifold 30, 39
   partial 30, 34 36
   invariant metrics on 30, 32 37
G-frame 47
Gauss-Bonnet-Chern theorem 61, 81
Gauss-Frenet map 57, 75
Gaussian curvature 61, 65, 69, 70, 77
Genus 61

Harr integral 10
Hodge operator 61, 65, 70, 77
Holomorphic curves 57, 67
   algebraic 79
   degenerate 64, 75, 84, 89
   horizontal 62, 64, 72, 84, 88
   projective 70, 85
   symplectic 72
Homogeneous space 27
Horizontal distribution 43
Horizontal submanifold 44

Induced representation 9
Inner product algebra 1, 2
Integrability condition 40
Integral element 96
Integral formula 61, 67, 78
Invariant 48, 49
Invariant complex structure 32, 47

Kähler form 80, 89

Laplace-Beltrami operator 61, 65, 77
Levi-Civita connection 65, 70, 77
Lie algebra 4
  Ad (G)-invariant inner product 10
  nilpotent 8
  radical of 8
  semisimple 8
  simple 8
  solvable 8
Linear isotropy representation 28
Linearly full 69

Maurer-Cartan form 36
  $(1, 0)$-components of 44
Maximal tori 12, 19
  integer lattice of 14
Moving frames 47, 68, 69

Normal form 100

Octanians 1, 2
$O(n; \mathbf{R})$ 3, 17, 21
Ordinary element 97
Orthogonal representation 9

p-forms 92
p-vectors 92
Pin (n) 3
Plücker embedding 82
Plücker formula 70, 79, 85
  symplectic 87
Positive root 23
Prolongation 99, 102
Pseudocomplex map 46

Quantization theorem 66
Quaternionic projective space 35
Quaternions 1, 2

Ramification index 84, 86
rank G 25
Regular element 97
Regular point 59, 63, 75
Roots 14, 20, 22

Schur's lemma 11
Simple root 23
Singular divisor 81
Singular metric 80, 86

Singular point 59, 63, 75, 79
SO (2*n*; R) -flag manifold 53
    partial 54
SO (*n*; R) 3, 16, 21, 24, 39
Sp (*n*) 3, 15, 21, 24
Spin (*n*) 3
Stiefel diagram 15, 18
Structure equations 36, 49, 54, 66
SU (*n*) 3, 15, 21
Subharmonic function 66, 77
Symmetric space 29
    exceptional 42
    hermitian 30, 42
    inner 29, 46
    types I, II 29

Tangent bundle 28

U (*n*) 3, 15, 21, 24, 39
U (*n*) -flag manifold 50
    partial 51
Unitary representation 9

Wang theorem 33
Weyl chamber 20
Weyl group 18